FRONT COVER
Referred to in the text as "Figure 1"

Passage from The Periodic Law

"The elements, if arranged according to their atomic weights, exhibit an evident periodicity of properties."
DIMITRI MENDELEEV, *1869*

to

A Periodic Table

The elements, if arranged in gapless vertical columns according to similarities* and in gapless rows according to their ordinal numbers in Mendeleev's arrangement yield a table with evident regularities.

*   Similarities of atomic structure [same number and same type(s) of valence-shell electrons]; also, similarities of properties as simple substances — *except in the case of light elements*, <u>especially hydrogen and helium.</u>

Legend

Blue	Halogen		F	Cl	Br	I	At	117
Green	Noble Gas		Ne	Ar	Kr	Xe	Rn	118
Yellow	Alkali Metal	Li	Na	K	Rb	Cs	Fr	119
Red	Alkaline Earth Metal	Be	Mg	Ca	Sr	Ba	Ra	120

Hatched downward to right	Coinage Metal		Cu	Ag	Au	111
Hatched downward to left	Volatile Metal		Zn	Cd	Hg	112

Black   Element with a maximum oxidation number of +3 following an element with a maximum oxidation number of +2

Gray   Element with a maximum oxidation number of +4 following an element with a maximum oxidation number of +3

The two lightest elements are located in the "Left-Step" Table so as to maximize its overall regularity regarding lengths of periods and columns.

The front cover's two diagrams are an answer to a physicist's question: "Does periodicity jump out at you?"

Periodic Table

Is	IIs	IIId	IVd	Vd	VId	VIId	VIIId	IXd	Xd	XId	XIId	IIIp	IVp	Vp	VIp	VIIp	VIIIp
1 H	2 He																
3 Li	4 Be											5 B	6 C	7 N	8 O	9 F	10 Ne
11 Na	12 Mg											13 Al	14 Si	15 P	16 S	17 Cl	18 Ar
19 K	20 Ca	21 Sc	22 Ti	23 V	34 Cr	35 Mn	36 Fe	37 Co	38 Ni	39 Cu	30 Zn	31 Ga	32 Ge	33 As	34 Se	35 Br	36 Kr
37 Rb	38 Sr	39 Y	40 Zr	41 Nb	42 Mo	43 Tc	44 Ru	45 Rh	46 Pd	47 Ag	48 Cd	49 In	50 Sn	51 Sb	52 Te	53 I	54 Xe
55 Cs	56 Ba	71 Lu	72 Hf	73 Ta	74 W	75 Re	76 Os	77 Ir	78 Pt	79 Au	80 Hg	81 Tl	82 Pb	83 Bi	84 Po	85 At	86 Rn
87 Fr	88 Ra	103 Lr	104 Rf	105 Ha	106 Sg	107 Bh	108 Hs	109 Mt	110	111	112	113	114	115	116	117	118
119	120																

IIIf	IVf	Vf	VIf	VIIf	VIIIf	IXf	Xf	XIf	XIIf	XIIIf	XIVf	XVf	XVIf
57 La	58 Ce	59 Pr	60 Nd	61 Pm	62 Sm	63 Eu	64 Gd	65 Tb	66 Dy	67 Ho	68 Er	69 Tm	70 Yb
89 Ac	90 Th	91 Pa	92 U	93 Np	94 Pu	95 Am	96 Cm	97 Bk	98 Cf	99 Es	100 Fm	101 Md	102 No

New Ideas in Chemistry
from
Fresh Energy for the Periodic Law

An Introduction to Leading Uses
of
The Left-Step Periodic Table

With a Foreword by Roald Hoffmann

Henry A. Bent

Bloomington, IN  Milton Keynes, UK
authorHOUSE®

AuthorHouse™
1663 Liberty Drive, Suite 200
Bloomington, IN 47403
www.authorhouse.com
Phone: 1-800-839-8640

AuthorHouse™ UK Ltd.
500 Avebury Boulevard
Central Milton Keynes, MK9 2BE
www.authorhouse.co.uk
Phone: 08001974150

First published by AuthorHouse 9/1/2006

ISBN: 1-4259-4862-6 (sc)

Library of Congress Control Number: 2006906617

Printed in the United States of America
Bloomington, Indiana

This book is printed on acid-free paper.

Our science is no better than our expression of it.

ORVILLE CHAPMAN

FOREWORD

From his first writings, a respect for the bounteous reality of the chemical universe coupled with the courageous drive to rethink every chemical observable have marked Henry Bent's work. His *Second Law* book came out just as I was beginning to teach introductory chemistry — through its simple language and realistic examples, it shocked me out of my conviction that thermodynamics could only be understood if one had control of those partial derivatives. His 1961 *Chemical Review* article traced more clearly than anyone had done previously the geometrical consequences of hybridization, and how geometry changes in turn affected hybridization. His 1968 *Chemical Review* article has a breathtakingly clear analysis (for its time) of how secondary interactions in the then just beginning prolific crystal structures could be made sense of as donor-acceptor interaction. The same article contains the first Bürgi-Dunitz diagram (for triiodide). Except it's by Bent.

The periodic table is our icon, and not for nothing, as they say. It remains a vital research instrument, for glassmaker and catalyst designer, for everyone who wants to ring the changes on the elements. It is also an intellectual tour-de-force, making sense of the atomic world. The robust utility of Mendeleyev's Table was already attested to by the 70 years that passed before this construction could be said to be deeply understood.

Henry Bent takes a fresh look at the innards of the table, and at its representations. The language is even and clear — Bent style. And his approach is obsessive — he returns again and again to the same points. But with not a touch of malice toward the prevalent illogic around our icon. That is also Bent. Even if one disagrees with him (as I do) on the problems that the ambiguity in the representations of the table pose for the community, one cannot but be impressed by the intellectual honesty, the feeling for the learner who is the reader, and for the <u>understanding</u> that permeate this truly original book.

Roald Hoffmann
*Frank H. T. Rhodes Professor
of Humane Letters,
Professor of Chemistry*
Cornell University
Nobel Laureate 1981

A Brief Dialogue with Roald Hoffmann

Alternative Title Pages for this Book
suggested by Hoffmann's Foreword

57 Reasons **(and counting) for**
Relocating Helium in Periodic Tables

Repeated Returns to The Helium Question:
"To Be or Not to Be?"

Henry **Be**nt

An Alternative to "Alternative Title Pages"
suggested by Hoffmann himself

The Fox and the Hedgehog
Loose in the Periodic Table

Working Titles

New Ideas

He/Be

f   d   p   s

SHORT ABSTRACT

Leading Uses of the Left-Step Periodic Table

- Direct tabular expression of the Periodic Law

- Point of departure for construction of less regular periodic tables.

- Identification of natural locations in periodic tables for the "problem elements".

- Graphic expression of block-to-block trends.

- Establishment of a correspondence with atomic physics.

- Suggestion of a physical explanation for the shapes of periodic tables.

SHORT ABSTRACT AMPLIFIED

Leading Uses of the Left-Step Periodic Table

- Direct tabular expression of the Periodic Law, by way of a set of natural Table Construction Conventions.

- Point of departure for construction of less regular periodic tables, by rearrangements of the Left-Step Table's blocks.

- Identification of natural locations in periodic tables for the "problem elements", particularly hydrogen and helium, placed at the top of the s-block.

- Graphic expression of block-to-block trends, particularly in (1) the distinctiveness of Groups' first elements, greatest in the s-block (with hydrogen above lithium and helium above beryllium), least in the f-block; and in (2) the similarities to each other of elements in blocks' first rows, least in the s-block (with helium adjacent to hydrogen), greatest in the f-block (with the "rare earths").

- Establishment of a correspondence between the integers of Chemical Periodicity and the quantum numbers of Atomic Physics. Ordinal numbers of blocks (arranged by size) are equal to the angular quantum numbers, ℓ, of the orbitals of blocks' predominant type of differentiating electrons in Bohr's Aufbau Process. Ordinal numbers of rows within blocks are equal to the orbitals' radial quantum numbers, r. An Index of Discontinuity, n, equal to the sum $r + \ell$, is an Orbital's Principal Quantum Number.

- Suggestion of a physical explanation for the leading feature of the shapes of periodic tables (late occupancy in Bohr's Aufbau Process of high-ℓ orbitals) in terms of an explanation for the coefficient "2" in the Madelung parameter $n + \ell$ written $r + 2\ell$.

ANOTHER ABSTRACT

Helium's placement above neon in periodic tables is a category mistake. The Periodic Classification of the Elements, according to Mendeleev, is a classification of *atoms,* not simple substances (such as "inert gases"). The most noble of the noble gases is not a Noble Gas. Much chemical evidence suggests that helium's natural position in periodic tables is, as physics suggests, above beryllium, where it supports a number of overlooked regularities in the Periodic System, including: Rules of Dyads, Triads, Group Sizes, Full Shells, First-Element Distinctiveness, and numerous other block-to-block trends; an isomorphism between integers generated by Chemical Periodicity and the quantum numbers of Atomic Physics, including a new Row-Orbital Correspondence; and unified plots of Secondary Periodicity. Discussed, also, are: an axiomatization of the Periodic System via a set of natural Periodic Table Construction Conventions for direct passage from the Periodic Law to the Left-Step Periodic Table; an explanation for late occupancy of high-ℓ orbitals in Bohr's Aufbau Process; chemical capture of the principal quantum number n; orbitals' nodal properties; new physical interpretations of the quantum numbers n and ℓ; natural and artificial classifications; classification of Groups by size; resolution of the zinc issue via a trend in ℓ-nobility; numerical indices of elements' distinctiveness; phenomenological definitions of different types of chemical kinships; the zinc-magnesium issue; locations in periodic tables of other "problem elements"; periodicity-like character of the order of atoms' excited states and the excited-state character of Chemical Periodicity; novel Mendeleevian "atomanalogies"; predictions based on the shape of the Left-Step Periodic Table; the Table's mathematical equivalence to Madelung's diagram; a new expression for Madelung's parameter $n + \ell$; a quantum-number/quantum-state spread sheet for study of the dependence of atomic orbitals' energies on their nodal properties; topological equivalence of Periodic Tables; Pauli's conceptual error in his induction of the Exclusion Principle from Periodicity's "magic numbers"; Natural Column Labels; novel, three-dimensional arrangements of atoms; and transformations of arguments for placing helium above neon in periodic tables into arguments for placing it above beryllium. Included are fifty-six, mostly entirely new figures, twenty-three appendices on special topics, and a number of historical, philosophical, and pedagogical perspectives.

List of Figures

Contents

xxiii

Introduction

"Chemical periodicity and the periodic table now find their natural interpretation in the detailed electronic structure of the atom" states a leading textbook of inorganic chemistry (1). Yet helium, an s^2 system, usually appears in periodic tables above neon, a p^6 system. Faced with that unique inconsistency in the Periodic System, users of periodic tables have several options.

Challenge the alleged inconsistency. "There's no problem. Helium and neon are inert gases (2)." Period. End of discussion.

Live with it. "Chemistry is complicated. One shouldn't try to over-simplify it" (2).

Dismiss it. "In chemistry chemical data trump physical facts" (2).

Or, one might try to domesticate the assignment by showing, e.g., that:

- Chemical routes to helium-above-beryllium exist.

- He/Be in periodic tables yields hitherto unnoticed regularities.

- Arguments for He/Ne are naïve, irrelevant, or incorrect.

This report (of results of a simple experiment, described shortly) addresses first the first two options (logically speaking, they're often interchangeable), then the third option. The strategy is to establish chemically what can be established chemically and then to introduce, by induction, connections with physics. Because the subject is the Periodic *System*, whose features are related to each other, the report's sections may be arranged in different logical orders, indicated, at times, parenthetically, by cross-referencing numbers in **boldface fonts**. Numbers in ordinary fonts within parentheses refer to an Annotated Bibliography, letters in parentheses to Notes (a).

An Added Note. Readers of a book on science, where, ideally, every feature has a logical reason behind it, might wonder: How does one rationalize use of page headers of *double lines* of *different thickness?* "When in doubt," said Mark Twain, "tell the truth." In truth, single lines looked attractive, initially, to outer and inner eyes: reminders of the horizontal lines of periodic tables' all-important periods — old hat, perhaps, but relevant, nonetheless. Seen subsequently in a book by William Jensen on Mendeleev and the Periodic Law were double-line page headers similar to those used herein. But for what *scientific* reason(s)? None came to mind, initially. "Sometimes the hardest things to see," say artists, "are the things that stare you in the face." What is this book's leading novel, utilitarian feature? Recognition, perhaps, of the support given to uses of the Left-Step Periodic Table in chemistry by the Mendeleev-Jensen Rule regarding a block-to-block trend in first-element distinctiveness (**30, 31**), exhibited graphically in periodic tables by *horizontal lines of variable thickness* (Figure 15). Omission in such instances of periodic tables' physically meaningless top horizontal lines yields short segments reminiscent of double-line page headers — and vice-versa.

He/Be: Arguments and Implications

1. Chemical Capture of the Left-Step Periodic Table (LSPT). All chemical routes to relocation of helium in periodic tables to a position above beryllium begin with graphic expression of the Periodic Law [top portion of Figure 1 (the front cover)].

With "Mendeleev's Line" in hand, the next step in domestication of He/Be lies in an answer to the question:

How does one go from Mendeleev's Line to a Periodic Table?

What, in other words, are one's Table Construction Conventions? Surprisingly absent in the history of Chemical Periodicity, given its centrality in inorganic chemistry, is explicit statement of specific conventions for construction of a periodic table from the Periodic Law. Cited sometimes are a desire for maximum exhibition of elements' similarities to each other (by way of adjacency) and a desire for maximum regularities in a table's shape. In fact, those two goals are, to some extent, as will be seen, mutually incompatible. One is led to try –

AN EXPERIMENT

What happens on application to Mendeleev's Line of the following conventions?

Periodic Table Construction Conventions

- Groups' members in vertical columns.
- Atomic numbers increase left-to-right, top-down.
- Gapless columns and periods.
- Maximum regularity in column and period lengths.

Reasons for the Construction Conventions

- *Verticality* for ease in scanning groups' members.
- *Sequentiality* of atomic numbers for graphic display of the Periodic Law.
- *Gaplessness* because gaps meant for Mendeleev missing elements.
- *Regularity* of column and period lengths for ease of description.

The Conventions work in concert. Particularly important is the Convention of Maximum Regularity. Most of Periodicity's overlooked features cited in this report are overlooked regularities.

Produced by the Conventions, unexpectedly, in view of the existence of some 700 periodic tables, is a *single* table: the helium-above-beryllium, 8-period, *s*-block-on-the-right, *fdps*, paired-periods, perfectly regular, periodic table, introduced by Janet in 1927 (3) and commonly called the Left-Step Periodic Table (bottom figure of the front cover). It's easily rearranged to less regular arrangements of its blocks, Figure 2 (next page).

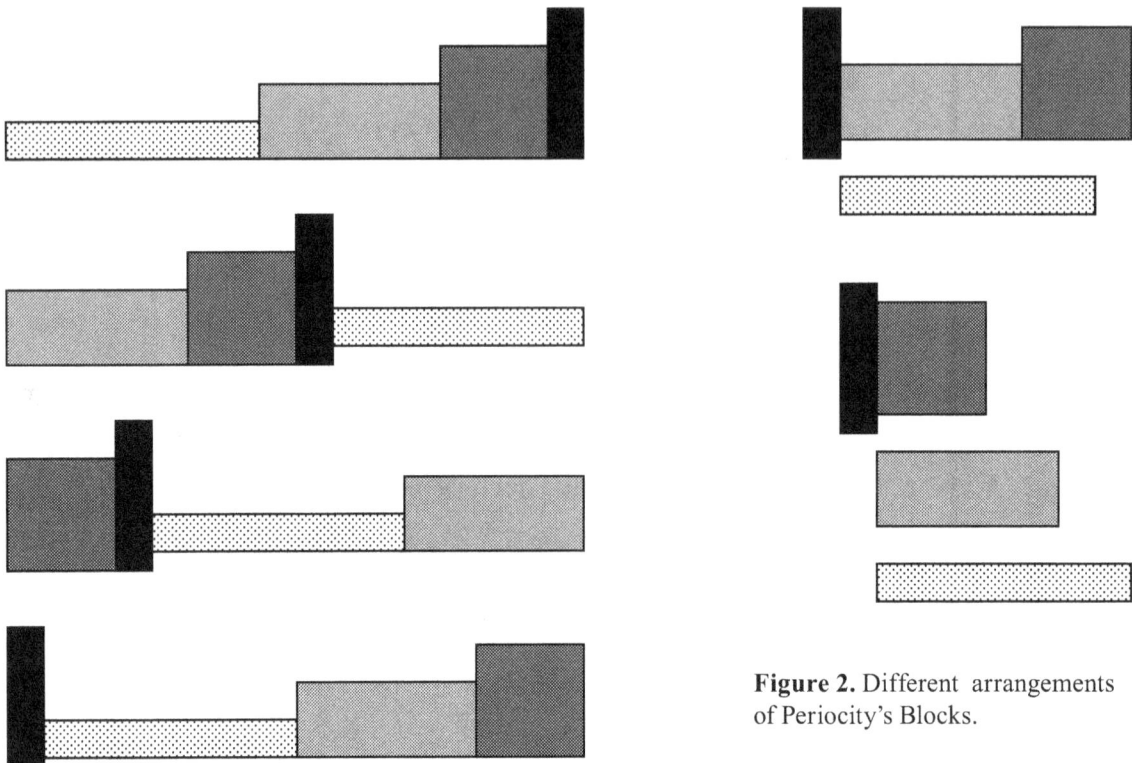

Figure 2. Different arrangements of Periocity's Blocks.

Figure 1 (the front cover) is the foundation of this report. It is called a *Chemical* Capture of the Left-Step Periodic Table because it is based on a chemical law, the Periodic Law. Not mentioned are electrons and orbitals (b).

2. Full Disclosure. "How did you arrive at your construction conventions?" asks a physicist. "By working backwards? By trial and error?" Actually, both. The Conventions were selected (i.e., induced) with the Left-Step Table in mind. Strictly speaking, therefore, they are, from an historical point of view, an *axiomatization* of the Table, not logical precursors to it. Their justification is the conventional justification for conventions in the inductive sciences: interesting implications.

3. Construction *Rules* for the Left-Step Periodic Table. Once the LSPT has been captured via a set of Construction *Convention*s, statement of a set of Construction *Rules* is relatively easy. *(I) Write down the number 2. (II) Beneath it in a vertical column list the atomic numbers of the alkaline earth metals. Include element 120.*

2

4

12

20

38

56

88

120

(III) To the left of each integer and down one level place the next integer, with space for intervening integers (between itself and the integer to its right). Those integers are 3 (after 2), 5 (after 4) . . . 89 (after 88), Figure 3.

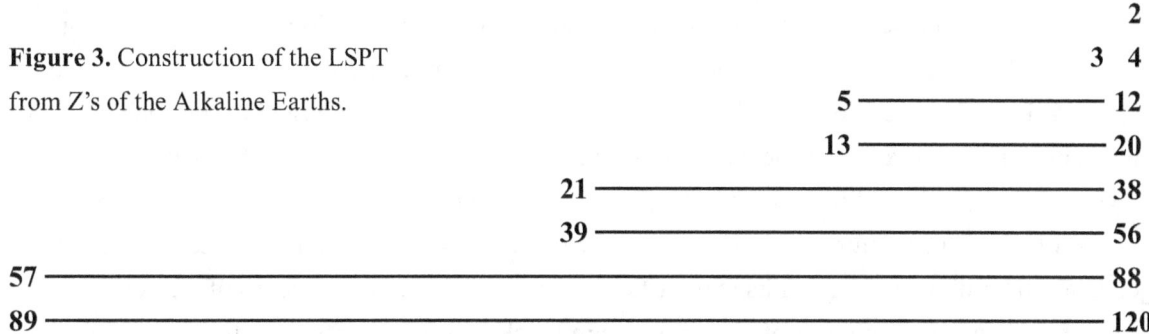

Figure 3. Construction of the LSPT from Z's of the Alkaline Earths.

(IV) Fill in missing atomic numbers. (V) Complete the array's paired-period character with a "1" immediately above the 3 (and to the left of 2).

To generate the initial set of integers, 2, 4, . . . 120, (I) write down the integers 1 2 3 4, each one twice; (II) square them; (III) double the squares; and (IV) add each double to the previous sum, starting with the first 2.

(I)		1	1	2	2	3	3	4	4
(II)		1	1	4	4	9	9	16	16
(III)		2	2	8	8	18	18	32	32
(IV)	0	2	4	12	20	38	56	88	120

4. Distinctive Features of the Left-Step Periodic Table. At first sight, it might seem that no periodic table could possibly be of less interest to chemists than the LSPT. In locating nature's most noble element at the head of a column of active metals and her most reactive metals adjacent to the Noble Gases, the LSPT violates twice over in, arguably, the most grievous ways possible, the familiar rule that adjacency in periodic tables implies similarity. [(Indignation regarding He/Be has been comparable to Henry Armstrong's indignation regarding the Bragg's model of sodium chloride (Appendix I)]. Transformation of the LSPT's two violations of the Similarity-Adjacency Rule into virtues is one of the leading goals of this report. The two disjunctions, it will be suggested, form, in fact, a harmonious whole.

The LSPT begins periods, except for its first two, in a curious manner, from the point of view of conventional, *sfdp*-type periodic tables. All of the LSPT's periods except its first two begin with an element whose maximum state of oxidation is III. That oddity is a reminder that –

Outer s-electrons are valence electrons in all blocks.

The requirement that periodic tables' periods begin with alkali metals does not follow, however, from the Periodic Law itself. It's an ad hoc — if, in some respects, a chemically useful — requirement that has delayed recognition of a large number of Periodicity's regularities.

The LSPT's most obvious distinctive feature to untutored eyes is its regularity. It is the only periodic table that combines verticality of congeners with gap-free periods and columns, in a regular manner. Because it is free of irregularities, it is a table of maximum periodicity: e.g., in number of paired-periods. No irregularities occur in lengths of its periods and columns (for Z through 120) — provided the Periodic System's leading "problem elements" (4), H and He, appear at the top of the Table. Proceeding from left-to-right, the Table has 4 steps, always upward, always of step height 2, and always of regularly decreasing depth: 14, 10, 6, 2. For comparison, the "long form" of the Conventional Periodic Table, helium above neon, lanthanides and actinides between the *s*- and *d*-blocks, has, left-to-right, 6 irregular steps (up, down, down, up, up, up) of irregular heights (7, 1, 4, 2, 2, 1) and irregular depths (1, 1, 14, 10, 5, 1).

As a table of maximum regularity, the LSPT is distinctive in the number of sets of integers it generates via ordinal numbers of its regularities' components: elements, groups of elements, blocks of groups of elements, columns of elements within blocks, rows of elements within blocks, rows of elements within the system, periods, and paired-periods (45).

Because the LSPT is unique in the simplicity of its construction from graphic expression of Mendeleev's statement of the Periodic Law, its most fundamental distinctive feature is, arguably, its relation to what might be considered to be, owing to its implications, a set of *natural* Construction Conventions (16). Particularly noteworthy is the Convention of Maximum Regularity. As its name suggests, the Convention reflects (through an induction to it, from the shape of the LSPT) and reveals (by way of deductions from it, through uses of the LSPT) Periodicity's *regularities*, the basis for calling Mendeleev's Classification of the Elements according to the Periodic Law "The Periodic *System*". The words "system", "periodic", and "regularity" in the phrases "Periodic System", "Periodic Law", and "Convention of Maximum Regularity" point to the same thing: a tabular expression of Chemical Periodicity of no irregularities.

Two particularly fundamental distinctive features of the LSPT are its concordance with (i) atomic physics (48) and with (ii) two of Mendeleev's leading principles regarding the Periodic Law (30, 34). Also distinctive and fundamental is the simplicity of an algebraic expression of the Table's form (49), which suggests, with a Row-Orbital Correspondence (50), a physical explanation for the shapes of periodic tables (61, 62, 64).

Possession of "distinctive" features by the LSPT implies, of course, a question: Distinctive compared to what? The Conventional Periodic Table? From a purely logical, if not historical, point of view, the relationship between the two tables is reciprocal. It raises the possibility of a gedanken reconstruction of events: *What would the logic of the situation have been had history been different?*

Suppose atomic spectroscopy, quantum mechanics, and the quantum numbers n, ℓ, and r had long been on the scene, together with all the lanthanides, the noble gases, and xenon tetroxide, when Mendeleev announced via verbal and graphic expression in the manner exhibited by the top part of Figure 1 (front cover) the existence of the phenomenon of chemical periodicity. What would Mendeleev's first tabular expression of the Periodic Law have looked like? Given his love of regularity, his distinction between natural and artificial classifications (20) [and, consequently, the possible irrelevance, in a natural classification of elements, of metallic and nonmetallic character (19)], and given his recognition of light-element distinctiveness (30), it seems plausible to suppose that one

6

of Mendeleev's first periodic tables might have been the LSPT. In that scenario the tables would have been turned, regarding the issue of conventionality. From the point of view of the Left-Step Table, today's Conventional Periodic Table has unconventional features: He above Ne; a footnoted *f*-step; and the s-step, in its height and width, out of step with respect to the other steps.

> *The shape of the Conventional Periodic Table*
> *is to some extent an historical accident.*

The author's physicist brother, Robert Bent, raises at this point a logical question: "Which table is the *conventional* one?" The Regular-Step, *fdps* Table? Or the Irregular-Step, *sfdp* Table? "The LSPT," he continues, "follows from a definite set of *conventions*. So it would seem that it should be called 'the conventional table'. You use the term 'conventional' to mean 'traditional' or 'widely accepted' or 'standard'. Should you use 'conventional' for the LSPT and 'standard' (or some other term) for the commonly used one?" Not necessarily.

In fact, both tables, *sfdp* and *fdps*, follow from a "definite [if unstated] set of conventions". For the traditional "conventional", *sfdp* table, those conventions are:

- o maximum oxidation number of +1 for elements of the table's first column on the left, reinforced by the convention that –

- o all metals be together in one part of the table;

- o lanthanides and actinides footnoted (owing, in part, to the belief that they are members of a single, super-group, "3d", or "3f" (**92, 94**)); and –

- o location of helium according to a rule of proximity for elements that as simple substances are similar to each other.

Those four conventions act in accordance with the meaning of an important derivative of the Indo-European root of the word "conventional": *prevent*. The conventional conventions, hardened to dogma, prevent one from recognizing many of Periodicity's regularities.

5. Row-Occupancy Rule. Owing to the LSPT's regularities, occupancy of its blocks' rows with increasing atomic number follows a simple rule. Call the ordinal numbers of its periods, starting at the top at 1, P and the ordinal numbers of its blocks, starting at the right at 0, ℓ, Figure 4.

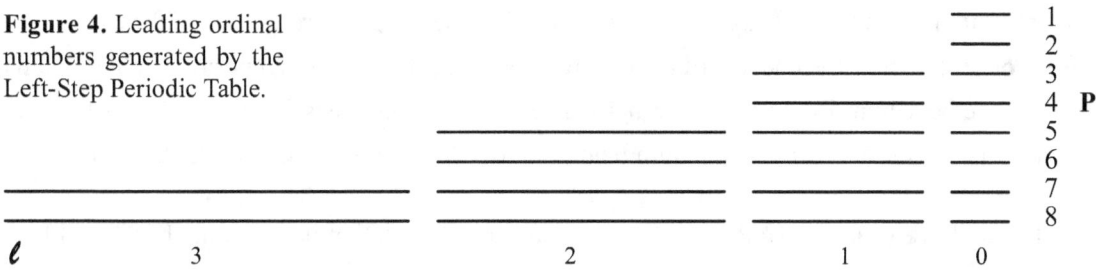

Figure 4. Leading ordinal numbers generated by the Left-Step Periodic Table.

P and ℓ are coordinates of blocks' rows. They are, in effect, the rows' chemical quantum numbers. The order of occupancy of rows with increasing atomic number may be simply stated:

Row-Occupancy Rule

Smallest **P** first and for a given **P** largest ℓ first.

For the "Long-Form" of the Conventional, *sfdp* Periodic Table the Row-Occupancy Rule is the more complicated statement: Smallest P_{sfdp} first and, for given P_{sfdp}, smallest ℓ first (namely, $\ell = 0$, of the *s*-block) and, thereafter, largest ℓ to smallest remaining ℓ (namely, $\ell = 1$, of the *p*-block). Illustrated, once again, is an arithmetical simplicity of the LSPT, immediately apparent to the outer eye.

6. General Implications of Chemical Capture of the LSPT. The construction of the LSPT sketched on the front cover has chemical, psychological, physical, philosophical, pedagogical, and practical implications. Because chemistry played a leading role in creation of Mendeleev's Line, which, with the Construction Conventions, yields the Left-Step Periodic Table, which has helium above beryllium, it seems reasonable to suggest that -

There exists a chemical route to the Left-Step Periodic Table and, hence, to location of helium above beryllium in periodic tables.

Recognition of a chemical provenance for The Left-Step Table, often dismissed by chemists as merely a "spectroscopic table" (useful, perhaps, as a mnemonic in assignment of electron configurations for atoms, but not otherwise significant for chemistry), has, itself, several implications. It provides psychological support for searches for additional connections between the Left-Step Table and Chemistry. It suggests possible existence of hitherto overlooked connections between geometrical features of periodic tables and atomic physics. It raises, as already noted (**2**), questions regarding the logical relation to each other of the Periodic Law, the Construction Conventions, and the Left-Step Table. It provides for the first time a simple yet rigorous route from the Periodic Law to periodic tables. And it raises four questions (to be discussed): (I) What is the physical significance of the "chemical quantum number" $P_{(fdps)}$ (**54**)? (II) Is it possible to generate from the Table an integer analogous to the principal quantum number *n* of atomic physics (**46**)? (III) Does selective relaxation of the Construction Conventions (other than the one regarding Maximum Regularity) yield other periodic tables that have helium above beryllium Be? And (IV) What logical justification exists for extrapolating from such periodic tables to the assertion that helium belongs above beryllium in *all* periodic tables (**100**)?

7. Logical Status of the LSPT–He/Be Connection. The relation between the LSPT and He/Be is synergistic. He/Be completes the LSPT's regularities. The LSPT's regularities demand — given the Convention of Maximum Regularity — He/Be. The argument from one to the other, some skeptics of the assignment He/Be have pointed out, is circular. Arguably, it must be.

Facts yield, by induction, theories, which, by deduction, yield back the facts (and, perhaps, new facts not identical with the initial facts but that, logically speaking, are essentially the same sort of facts, called "predictions"). The LSPT yields, by induction, the Convention of Maximum Regularity, which yields back, from the LSPT, sans He, the prediction that such an element should exist, above Be. The charge of "circularity" indicates occurrence in the argument of an induction, expressed, in the present instance, by the Convention of Maximum Regularity.

A striking example of circularity in the inductive sciences is the statement that "Periodic tables are based on two of the leading principles of modern physics: the laws of quantum mechanics [which account, in part, for occurrence of the factor $(2\ell + 1)$ in the expression $2(2\ell + 1)$ for the lengths of blocks' rows in periodic tables] and Pauli's Exclusion Principle". In truth, however, Pauli's Principle was, in fact, an induction, by Pauli, based on the lengths of periodic tables' periods (Appendix II).

The circularity of the LSPT-He/Be relation is analogous to the circularity of the relation between the Periodic System, which Mendeleev based on atomic weights, and atomic weights, which Mendeleev corrected, via use of the Periodic System. Employed in both instances is a bootstrap circularity to cope with a Catch-22: understanding X requires Y, which requires, for its understanding, X. Hence, frequent occurrence in this report (vide infra) of references in a Section X to a Section Y, Y > X (c).

8. Global and Literal "Atomanalogies" Involving Helium. Circularity of the argument for placement of helium in periodic tables via use of the Convention of Maximum Regularity arises, in part, from the fact that the problem-element problem is the inverse of Mendeleev's gap-problem. In the gap-problem, Mendeleev had locations for his "eka-elements", but not the elements themselves. In the problem-element problem, the elements are at hand, but not universally agreed-upon locations for them.

In predicting existence of gap-filling elements, and what their properties might be, Mendeleev relied on a Principle of Regularity *on a Local Scale*, in the form of what he called "atomanalogies": namely, interpolations involving properties of elements immediately before and after and above and below gaps in his periodic tables (5). Proposed locations for hydrogen and helium are, of course, not so bounded. The Periodic System's first two elements lie on the *edges* of periodic tables and not, like Mendeleev's "eka-elements", *within* the tables. Locating H and He in periodic tables by Mendeleev's method requires, consequently, use of the Principle of Regularity *on a Global Scale*, in the form of a periodic table in which each element contributes to the table's *overall regularities*, which, in turn, determine placement of each element.

The overall shape of the Left-Step Periodic Table is a powerful tool for locating the "problem elements" in a classification of the elements according to the character of Mendeleev's Line. On reflection, that's not surprising. The Table follows directly from the Line (Figure 1). Its success rate in addressing the challenges raised by the "problem elements" H and He, La and Ac, Lu and No, and Mg and Zn is 100 percent.

The situation is similar to a "very beautiful and simple explanation" of possibly "the most distorted [spectroscopic] series known. . . [I]t was not until the spectrum had been thoroughly analyzed and everything else accounted for, that the higher members of the series were definitely identified" (6).

It's easy to place the last piece in a jigsaw puzzle — or to predict, at that stage, that a piece is missing. Similarly, given a complete LSPT except for location of its first two members, even young students who've no knowledge of chemistry usually see quickly where to place the Periodic System's two lightest elements, in accordance with the logic of the LSPT — i.e., in accordance with an Assumption of Maximum Regularity.

In addition to global (full-table) atomanalogies, discussed throughout this report, the LSPT's first two periods

$$H \quad He$$
$$Li \quad Be$$

suggest a literal analogy:

$$H : Li :: He : Be$$

Replacing the "is" of ":" by "/" and the "as" of "::" by "=" yields the expression –

$$H/Li = He/Be$$

Solving for "He" yields -

$$He = Be (H/Li)$$

Insertion on the right-hand-side of the atoms' first-stage ionization energies yields –

$$He = 9.32 \text{ eV} (13.6/5.39) = \textbf{23.5 eV} \quad \text{Observed: } \textbf{24.6 eV}$$

Additional "atomanalogies" involving helium suggested by the shape of the LSPT are cited in Appendix III.

9. Minimum Chemical Data Required for Construction of the LSPT and Anxiety of Influence.

Because the LSPT has no irregularities in the lengths of its periods and columns (for Z_{max} equal to 120), the chemical data highlighted in Figure 1 over-determine the Table. Sufficient for its construction are locations in Mendeleev's Line of the alkaline earth metals (three members of which constituted Dobereiner's first triad) and members of one other Group. Distances of that Group's members to the next alkaline earth metal are the same for all members of the Group — but not distances, going forward with atomic numbers, from the alkaline earth metals to members of the Group in question. Consequently, *in the absence of gaps within periods* (third Construction Convention), the Group's members, if arrayed in a *vertical column* (First Convention), must lie to the *left* of the alkaline earth metals. That's true of any group (in accordance with the Convention of Maximum Regularity).

Locations in Mendeleev's Line of the alkaline earth metals and members of any other Group determine, by way of the formative Construction Conventions, membership of *all* of Periodicity's other Groups. Such a wholesale return of results from so skimpy a statement of facts might seem, initially,

10

highly improbable. In fact, however, Mandeleev's Line is one of science's leading graphic expressions of information. Latent in it are all the features regarding Chemical Periodicity reported in this report.

With helium above beryllium, most of Periodicity's newly recognized features can be gleaned from the Conventional Periodic Table, for it is similar to the LSPT in most respects. Indeed, all things considered, the two tables are nearly identical! Regarding elements' assignments to Groups, the two Tables agree in over ninety-nine percent of the cases. In only one instance do they disagree. Agreement regarding Groups' assignments to blocks is one hundred percent. Agreement regarding Groups' arrangements within blocks is one hundred percent. Agreement regarding ordinal numbers of blocks' rows with increasing atomic numbers is one hundred percent.

With so many points of agreement between the LSPT and the Conventional Periodic Table, which, indisputably, is a chemical table, particularly in its location of the *s*-block, it would seem remarkable, indeed, in retrospect, if, in fact, the Left-Step Periodic Table hadn't many connections with Chemistry, beyond those cited in the previous paragraph.

Unlike the LSPT's many overlooked connections with Chemistry, described herein, chemists have not overlooked, entirely, the LSPT's connections with Physics. [Placement of helium above neon does negate a connection between ordinal numbers of blocks' rows and numbers of orbitals' radial nodal surfaces (**48**) for the neon and beryllium columns of periodic tables and, hence, destroys the standing of that connection as a universal rule.] By labeling the LSPT a "spectroscopic table", or "orbital filling chart", and not a "true" periodic table, chemists have, however, implied that connections between the LSPT and physics are, for the most part, for chemistry, unimportant.

"For a physicist," writes Robert Bent, "connections with atomic physics are what make the LSPT 'right'. . . It seems strange indeed that some chemists don't consider concordance with electron configurations to be an outstanding feature of the LSPT." They do — but consider that to be the end of the LSPT story. Its concordance with physics, according to conventional wisdom, is its *only* outstanding feature.

Chemists' lack of interest in the LSPT may arise, in part, from what literary critics call "an anxiety of influence": of atomic physics on chemistry. A chemist is reluctant to see the hard won advances in chemical knowledge summarized in the concepts "element", "families of elements", "atomic weights", and the "Periodic Law" replaced by rote, arithmetical manipulation of the progression 1 1 2 2 3 3 4 4 (as in Section **3**). "It can't be that simple," one's tempted to say. And, of course, it wasn't. That's progress. Hence the fear, however, of creating young chemists-in-the-making who may be able to write down the electronic structure of Br_2, but who do not know the difference between the simple substance "bromine", a dark, red-brown, low-boiling, corrosive, highly toxic liquid, shipped in glass bottles, and packing-case vermiculite.

10. Questions for Theoreticians. Fruitful theories, it's said, raise more questions than they answer. (Sometimes that essential feature of a theory-in-evolution is taken by a theory's opponents to be a fatal flaw.) Section **9**'s title raises several questions.

Are *all* the cited data necessary? Might the Table Construction Conventions of Figure 1 and only *one* Group's atomic numbers, 2 4 12 20 38 56 88 120, yield a single table? If so, what's implied? That there's something distinctive about the atomic numbers of the alkaline earth metals? If so, what does that distinctive character imply regarding the character of a physical theory of the Left-Step Table's shape? Is there a simple physical interpretation of the "magic numbers" generated by differences in the atomic numbers of the alkaline earth metals: 8 8 18 18 32 32?

Our friendly physicist adds this question: "Knowing what we do today it seems natural to expect <u>no gaps</u>, but I wonder why Mendeleev thought this had to be so. Was this a guess, or was there some compelling reason in his day for believing this? Did he believe that nature abhorred gaps? I can see why chemists might, but why should nature? Does this mean man is special in the overall scheme of things after all — that nature was created to please man with regularities? Or was it the other way around, that man was created [by evolution] to appreciate [for purposes of survival] nature's regularities? What if it had turned out that a regular periodic table was not possible? Would there then have been no atomic physics as we know it today? It's very mysterious how everything fits together, and almost makes one believe in creationism. . ." Also: "Why is 'Maximum Regularity' a *natural* assumption? Why do we (and/or nature) care about regularity? That regularity should exist is a religious-like faith, it seems to me. There could have been just chaos, but then I suppose, we would not exist to wonder about these things."

A logician might wonder: Because Mendeleev's famous predictions regarding "eka" elements were largely based on the Periodic System's first two Newland "Octaves", Li Be B C N O F and Na Mg Al Si P S Cl, wasn't his assumption of no gaps wihin his tables' series rather like assuming that, since the first 99 numbers are less than 100, all numbers are less than 100? Actually, Mendeleev's assumption was more like assuming that, since there are no gaps in the sequence of the natural numbers, there are no gaps in the natural sequence of the elements in Mendeleev's Line, Appendix IV.

11. A Maximum-Oxidation-Number Route to the Left-Step Periodic Table. The route to the LSPT pictured in Figure 1 does not depend on knowledge of electrons, orbitals, quantum numbers, an orbital-occupancy rule, and Pauli's Exclusion Principle. Required, however, is knowledge that: (i) Be, although rather different from members of Dobereiner's leading triad Ca-Sr-Ba, is, in fact, in their Group; (ii) Mg, historically often grouped with Zn, Cd, and Hg, is, in fact, in the Ca-Sr-Ba Group; and (iii) the distinctive elements Cu, Ag, and Au are, in fact, congeners. One might wonder:

> *Is it possible to capture the Left-Step Periodic Table from chemical data without knowledge of Group-membership for any elements whatsoever?*

Consider, in the format of Mendeleev's Line (the top portion of Figure 1), elements' maximum oxidation numbers, Figure 5 (immediately below).

1 0 1 2 3 4 5 2 0 0 1 2 3 4 5 6 7 0 1 2 3 4 5 6 7 6 5 4 4 2 3 4

5 6 7 2 1 2 3 4 5 6 7 8 6 4 3 2 3 4 5 6 7 8 1 2 3 4 4 3 3 3 3 3

4 3 3 3 3 3 **3 4 5 6 7 8** 6 6 5 3 3 **3 4 5 6 7 8** 1 2 3 4 5 6 7 7 6

4 4 4 3 3 3 3 3 4

In bold face in Figure 5 are ascending sequences of maximum oxidation numbers, MON. In such sequences, no integers are missing. Alignment at the left of all **1**'s (the alkali metals) yields Figure 6.

Figure 6

1 0
1 2 3 4 5 2 0 0
1 2 3 4 5 6 7 0
1 2 3 4 5 6 7 6 5 4 4 **2 3 4 5 6** 7 2
1 2 3 4 5 6 7 **8** 6 4 3 **2 3 4 5 6 7 8**
1 2 3 4 4 3 3 3 3 3 4 3 3 3 3 3 **3 4 5 6 7 8** 6 6 5 **2 3 4 5 6 7 8**
1 2 3 4 5 6 7 7 6 4 4 4 3 3 3 3 3 4

All bold-face sequences are aligned vertically in Figure **6** *except the two sequences* **3 4 5 6 7 8** in the next-to-last line. Vertical alignment *at the right* of *all bold-face* **2**s aligns vertically *all* bold-face sequences, Figure 7.

Figure 7

1 0
1 2
3 4 5 2 0 0 1 2
3 4 5 6 7 0 1 2
3 4 5 6 7 6 5 4 4 2 3 4 5 6 7 2 1 2
3 4 5 6 7 8 6 4 3 2 3 4 5 6 7 8 1 2
3 4 4 3 3 3 3 3 4 3 3 3 3 3 3 4 5 6 7 8 6 6 5 2 3 4 5 6 7 8 1 2
3 4 5 6 7 7 6 4 4 4 3 3 3 3 3 4

One might say, in the words of Mendeleev, "the forms of oxides and . . . atomic weights . . . [give] us the means to erect an unarbitrary system as complete as possible" (5).

Boxed in Figure 7 are *maximum oxidation numbers equal to or less than those immediately to their left within their blocks*. Helium stands out as, arguably, the most distinctive boxed element in the Periodic System.

Let **X** stand for an oxidation number within a box. Then one may say that:

- No block starts with an **X**.
- Within blocks **X**'s appear rightward and disappear downward.
- Once **X** disappears downward in a column, **X**s do not reappear.
- Once started in a block's row, **X**s continue to the end of the row (except in the case of Tb in the lanthanide series).
- Relative numbers of **X**'s within blocks decrease in the order

$$f > d >> p >>> s$$

A block-to-block trend exists, also, in the number of first-row elements whose maximum oxidation numbers are less than their Groups' maximum oxidation numbers.

s-block:	He
p-block:	O F Ne
d-block:	Fe Co Ni Cu (Zn)
f-block:	*Pr Nd Pm Sm Eu Gd Dy*

Numbers of exceptions, 1, 3, 5, 7 (provided Zn is included among the exceptions, based on reported Hg(III)), are equal to $2\ell + 1$.

12. A Caveat and Alternate Chemical Route to the LSPT and He/Be. The maximum-oxidation-number route to the LSPT and He/Be described in Section **11** is, *strictly speaking,* not another, *independent*, chemical route to that destination, inasmuch as Mendeleev's statement of the Periodic Law, upon which the route displayed in Figure 1 is based, is, itself, based, in important measure, on maximum states of oxidation (hence, e.g., Mendeleev's placement of hydrogen in the same Group as the alkali metals, and carbon in the same Group as lead). Described below is an independent route to He/Be.

The Generic Mendeleevian Line of Figure 1 exhibits an incomplete progression in *differences in ordinal number*s *of* s*uccessive alkaline earth metal*s. Starting at the left, with beryllium, those differences are: 8 8 18 18 32 32. Division by 2 yields 4 4 9 9 16 16. Square roots are 2 2 3 3 4 4. Missing is an initial 1 1, rectified by considering He to be the first member of the alkaline earth metals Group, and counting the distance to it from the beginning of Mendeleev's' line as the first "2". That rectification of the "magic" differences 2, 8, 18, 32 by inclusion of an initial 2 amounts to considering He along with the alkaline earth metals to be termini of periods of a periodic table, completed as in Section **3**.

13. Two Classes of Blocks. Groups' maximum oxidation numbers, Gmax, define two classes of blocks. There are blocks of groups, call them type A, in which numbers of valence-shell electrons never exceed Gmax. And there are blocks of Groups, call them type B, in which numbers of valence-shell electrons sometimes exceed Gmax. The *s*- and *p*-blocks are type A. The *d*- and *f*-blocks are type B. Orbitals of differentiating electrons of elements in A blocks are "outer" orbitals, those of B blocks "inner" orbitals.

The block designators A and B correspond to suffixes A and B in the American ABA column-labeling scheme, Figure 41 (except for the case of the unlabeled *f*-block). Type A blocks include all, and only, "main groups". Type B blocks include all "transition metal groups".

The LSPT, criticized for not being like the Conventional Periodic Table in locating all metals together, does locate all type A blocks together and all type B blocks together, in the order, right-to-left, A A B B. In the Conventional Periodic Table the order is, either direction, A B B A.

14. Physical Routes to the LSPT. A plot, downward, of increasing values for atomic orbitals of P = $n + \ell$ against, leftward, increasing values of ℓ with, owing to orbital degeneracy, spin, and the

Exclusion Principle, $2(2\ell + 1)$ entries for each allowed set of values of $\ell \le n - 1$ $[= (P - \ell)/2]$ yields, when read for increasing values of Z, left-to-right, top-down, the Left-Step Periodic Table, Figure 4. One might wonder, therefore: Is it possible to capture the LSPT *purely from physical data*, in a manner analogous to that used with maximum oxidation numbers, without reference to orbitals, quantum numbers, the Pauli Exclusion Principle, and Madelung's orbital occupancy rule (smallest $n + \ell$ first and, for given $n + \ell$, largest ℓ first)?

Replacement in Mendeleev's Line of maximum oxidation numbers by atoms' radii (following in the footsteps of Lothar Meyer and his famous atomic volume plot) or by their first stage ionization energies, and then vertical alignment (with appropriate cuts) of the Lines' prominent peaks yields the Left-Step Periodic Table (7).

15. Visible Spectra of He, Hg, H_2, and Ne. Yet another route through physics to He/Be is experimental atomic spectroscopy. One senses immediately the import of the voluminous physical data in support of the empirical classifications of He and Ne as s^2 and p^6 systems, respectively, on viewing through a student spectroscope the line-lean visible part of the spark spectra of s^2 helium, mercury, and hydrogen, compared to the rich spectrum of p^6 neon.

Existence of several routes to the LSPT supports the idea that it is a *natural* arrangement of the elements.

16. Natural Classifications. For a classification to be deemed natural, stated the polymath William Whewell in 1857 (8), it must [like dependable calculations and reliable experimental results] be arrived at by [at least] two [logically independent] routes.

<u>Whewell's Criterion for Naturalness</u>
The arrangement obtained one way must coincide
with the arrangement obtained another way.

According to Whewell's criterion, existence of different routes to the Left-Step Periodic Table supports the assertion that –

> The Left-Step Periodic Table is one of the leading *natural* arrangements of physical entities in the history of human thought.

17. Bayesian Logic and the Probability of the LSPT's Utility in Chemistry. "The main sticking point for chemists in your discussion of Periodicity," wrote professor Emil Slowinski of an early draft of this report, "seems to be location of helium above beryllium."

> "The noble gas helium [according to its electron configuration] should be placed above beryllium, but *no form of the periodic table places it so* because it is clearly a noble gas and *must* go above neon even though this has an electron configuration of a filled *p*-shell" (4), emphasis added.

"[T]ransfer of He from the inert gases to the alkaline earth metal group [as the LSPT suggests] is so radical that it is not recommended" (9).

"[H]elium's uniqueness places it among the noble gases rather than the alkaline earth metals" (10).

Even an advocate of the LSPT regards it to be "An Arithmetical Table", with "no [chemical] justification for grouping helium with [the alkaline earth metals'] column" (9). It's no surprise, therefore, that the LSPT has not been seen to be, also, as we've seen, or will see: a table that can be based on chemical data in several ways; a way to settle locations in periodic tables of the problem elements; a source of a Row-and-Orbital-Occupancy Rule; a Meta-Trend Table; the Simplest Route from the Periodic Law to the Conventional Periodic Table; a way to rationalize the zinc issue; a source of natural column labels; a source of new ordinal numbers of periods for plots illustrative of secondary periodicity; and, generally, a source of new ideas for inorganic chemistry through fresh energy for the Periodic Law.

"A nice list of chemical things the LSPT does!" writes our friendly physicist. "So, the LSPT has dual merits: it does nice things for pure chemistry while being in concordance [vide infra] with modern physics. It unites two fields."

The presence of a heavy horizontal line *within a column* of Simmon's LSPT (9), beneath H and He, is the first instance of such an addition to the LSPT (enlarged upon in Section **18**). It is a hybrid of two views regarding H and He and periodic tables. One of those views views H and He as being *quantitatively* different from other elements (Simmons), as, indeed, they are, in the sense that, unlike other atoms, atoms of H and He have no inner-shell electrons. Expressive of that view in periodic tables is location of H and He in a separate "block" above the other elements. The other view of H and He views them as being *qualitatively* different from other elements (this report), in the sense of being *the leading instances of First-Element Distinctive*ness (**30, 31**). Expressive of that view is location of H and He as leading elements of the *s-block*.

The question Which view of H and He is "correct"? is a physically incorrect question. It's like asking: Which view of a house is "correct"? A front view? Or a side view? Or: Which theory of the electronic structure of matter is "correct"? Molecular orbital theory? Or valence bond theory? In both instances both views are useful, in different ways. An impartial observer holds in mind, simultaneously, all partial views (pun intended). The truth is the whole truth, not merely one part of it. At our comprehending best, we're all cubists, so to speak, in thought, if not on paper. This report is, for the most part, a report of the multidimensional character of Chemical Periodicity projected onto two dimensions, in the form of periodic tables. Discussed in Section **98** are three dimensional arrangements.

A second sticking point for chemists regarding this report is the LSPT's second distinctive feature: location of the *s*-block on the right-hand-side of a periodic table. Wrote a noted inorganic chemist in 2000 regarding an early version of this report, titled "*f d p s*":

"I have a strong objection to a Left Step Table, because it does not reflect the smooth representation of electronegativity, which is a great merit of the present organization. An important corollary to this is the logical presentation of the periodic trend in acid or base properties of the elements' oxides in water."

Called to mind by objections to fdps and He/Be is a remark variously attributed to Bohr, Pauli, and Einstein:

"Is it crazy enough to be right?"

If by "right" one means "particularly useful", then the answer to the BPE question regarding block order *fdps* and helium-location He/Be, suggests this report, is Yes. On reflection, perhaps that's not surprising, since, according to Bayesian logic, statements stand chances of being particularly useful if deemed at the outset to be particularly outlandish. Fools persisting in their folly, says the old Chinese proverb, find wisdom — or, at least, in the present instance, in studies of the implications of *fdps* and He/Be, *surpri*ses, beginning (truth be told) with helium above beryllium (instead of above neon, where, for almost half a century, the author had supposed it belonged) and continuing through the remainder of the conclusions reported in this report.

18. Graphic Display in the Format of the LSPT of Block-to-Block Discontinuities. The LSPT's unconventional block order fails to exhibit along its periods the famous trend along conventional periods, from metals to nonmetals, reducing agents to oxidizing agents, and basic oxides to acidic oxides (Noble Gases excepted). In fact, those trends are not entirely "smooth" (d). Nonetheless, replacement of the (somewhat) irregular trends by *major discontinuitie*s in elements' properties *within* the *fdps* periods, on exiting the *p*-block for the *s*-block, is a jolt, particularly in the absence in the LSPT of any graphic indications whatsoever (such as the period breaks between *p*- and *s*-blocks in conventional periods) of that striking feature of descriptive inorganic chemistry. That characteristic shortcoming of the LSPT, alone, is often assumed to mean, naturally, that a periodic table that conceals one of the leading features of the chemistry of the elements is unlikely to be a useful table in chemistry. The striking chemical discontinuity on entering the *s*-block — from itself or, as is usually the case, from the *p*-block — can be indicated, if less dramatically than usual, by inserting in the LSPT between the *s*- and *p*-blocks, in the spirit of Simmons' line beneath H and He in his LSPT, a heavy vertical line, Figure 8, below.

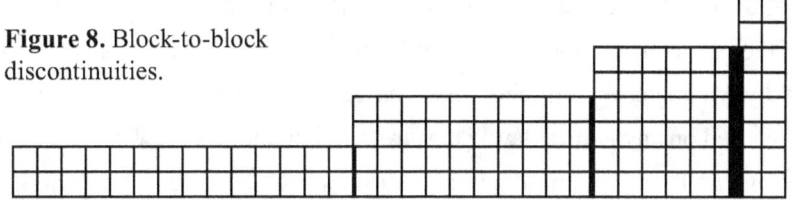

Figure 8. Block-to-block discontinuities.

Progressively lighter vertical lines in Figure 8 above between the *p*-block and the *d*-block and between the *d*-block and the *f*-block indicate that the discontinuity in going from ytterbium, at the end of the first row of the *f*-block, to lutetium, at the beginning of the third row of the *d*-block [Δ(IE) = -0.83 eV], is less [by about a factor of four] than the discontinuity in going from zinc, at the end of the first row of the *d*-block, to gallium, at the beginning of the third row of the *p*-block [Δ(IE) = -3.4 eV], which is less [by about a factor of five] than the discontinuity in going from neon, at the end of the first row of the *p*-block, to sodium, at the beginning of the third row of the *s*-block [Δ(IE) = -16.5 eV]. Transformed is a defect into a virtue. Prepared is the LSPT for its transformation into the Conventional Periodic Table (by movement of the *s*-block to the left). And illustrated is one of the leading uses of the LSPT: as a ground for display of block-to-block trends (**91**).

19. Metals, Nonmetals, and the Periodic System. The requirement that metals and nonmetals occupy distinct regions in Periodic Tables (hydrogen excepted?) is not part of the Periodic Law. It's a separately imposed, arbitrary requirement. In Mendeleev's famous "short-form" periodic table, first exhibited in 1870, metals and nonmetals intermingle, somewhat. Earlier, in 1854, the American chemist Josiah P. Cooke had remarked that –

> "Phosphorus and sulphur, which are not chemically allied, are frequently placed consecutively, while arsenic, antimony, and bismuth, in spite of their close analogies with phosphorus, are described in a different part of elementary textbooks on chemistry. *This confusion,* which is a source of great difficulty to the learner, *arises in part from retaining the <u>artificial</u> cla*ssification *of element*s *into metal*s *and metalloid*s. . *."* (11), emphasis added.

Similarly, Newlands wrote in 1863, in an article on "*natural* groups", that "in the difficult problem of grouping elementary bodies" he had taken "*no notice of the ordinary distinction between metal*s *and nonmetallic*s" (12), emphasis added.

One of the most famous trends in the Periodic System is the trend downward in Groups toward increasing metallic character. For the super-metals of the *s*-block, the metal-nonmetal transition occurs sharply, immediately after the two lightest *s*-elements. It occurs less sharply in the *p*-block, and disappears in the *d*- and *f*-blocks, which contain no nonmetals.

Although the LSPT does not group all the metals together, it does group together, on the right, all the commonly named Groups and all the gaseous elements, including Nature's four most volatile elements, *in order* (left-to-right) *of increasing volatility:* F, Ne, H, He.

20. Artificial and Natural Classifications. Classification of helium with neon in the Periodic Classification of the Elements combines in a single table two different types of classifications of the elements, called by Mendeleev "artificial" and "natural". Dualistic classifications, such as metal-nonmetal, are not based on any natural law. They are, in Mendeleev's view, "artificial" classifications, *compared* to the one and only "natural" classification of the elements, The Periodic Classification, based on the Periodic Law (Appendix V).

Dualistic, either-or classifications of the elements are based on a single, arbitrarily selected property. Because elements, as simple substances, have many properties, they can be classified, artificially, in many ways, according, e.g., to density, volatility, hardness, abundance, toxicity, metallic character, reactivity, and so forth. Useful though such classifications are in particular instances, they do not lead beyond themselves, to broad generalizations. Consequently, they do not lead to a *unique,* natural classification. In part that's because substance's properties are "functions of state".

The condition of metallicity, for instance, like an element's physical state, is contingent on the environment. At high pressures, all elements are metallic. At high temperatures and/or very low pressures, all elements are nonmetallic gases. At intermediate temperatures and pressures, some elements are metallic, some nonmetallic. The Periodic Law holds, however, at all temperatures and pressures for which the elements' atoms are distinct individuals, regardless of the elements' physical states and electrical conductivities.

Admittedly, the Table Construction Conventions cited at the outset for Figure 1 (**1**) are, like conventions regarding locations of metals in periodic tables, not part of the Periodic Law. Judged, however, by their consequences (the LSPT, and its implications), they are not arbitrary.

21. Construction of the Conventional Periodic Table from Chemical Data. Among arguments for considering the LSPT to be more than merely a "spectroscopic table", of interest to physicists, perhaps, but not to chemists and chemical educators, is the fact that it's easy to pass from the LSPT to tables of less regularity, including chemistry's Conventional Periodic Table, Figure 2, top right.

The Left-Step Periodic Table is merely a step away from the "long form" of the Conventional Periodic Table (Figure 2, bottom left).

A Simple Passage

> The most direct route from the Periodic Law to the Conventional Periodic Table passes through the Left-Step Periodic Table.

The LSPT is not the only periodic chemistry students should see — only, perhaps, the first one.

If the Conventional Periodic Table is considered to be a chemical table, then, too, the LSPT might be considered to be, by lineage, a chemical table, in rectified form in order to reveal global regularities concealed by the irregular shapes of less regular periodic tables.

The four periodic tables on the left of Figure 2 may be designated, downward, as the *fdps, dpsf, psfd,* and *sfdp* tables. Mazurs (3) attributes the *fdps,* left-step table to Janet (1927), the *dpsf* table to Mendeleev (1869, updated), the *psfd* table to Schmidt (1911), and the *sfdp* table to Werner (1905). Mendeleev's (gapless) *dpsf* table features the "transition metals" on the left. (Mendeleev's table would have gaps, however, were it extended to the *g*-block.) Schmidt's *psfd* table features Mendeleev's "typical elements" on the left and the volatile metals at the far right. The double-footnoted table at the bottom right is a Sanderson-type Table. It is the only table that (appears) to place the *s*-block (black) immediately to the left of the *p*-block (gray).

In the first two tables (those of Janet and Mendeleev), metals of the *s*-block (black) are separated from metals of the *d*-block (light gray) by nonmetals (in the dark gray block). In the table at the lower left [the so-called "long" form of the Conventional Periodic Table (upper right)], the two "Main Groups" of the Periodic System that comprise the black block (the alkali and alkaline earth metals) are separated from the System's other six "Main Groups", which comprise the *p*-block (gray), by metals of the *d*- and *f*-block. The Conventional Periodic Table, in summary, groups all metals together, at the expense of not grouping all "Main Groups" together. The Left-Step Table, on the other hand, groups all "Main Groups" together, at the expense of not grouping all metals together.

22. "Transition Metals": an Anachronism, Today? The LSPT exhibits Main Groups of the *p*-block adjacent to Main Groups of the *s*-block. Absent in the LSPT is the original reason for considering all elements of the *d*- and *f*-blocks to be "*transition* elements", in transition between Main Group elements of the *s*-block and Main Group elements of the *p*-block. (Mendeleev's "transition elements" were those of the *d*-block in transition between maximum oxidation numbers of the extended iron Group, of 6 or 8, to the 2 of the zinc Group.)

Redefinition of the phrase "transition elements" in terms of distinctive properties of the majority of *d*- and *f*-block elements raises the frequently asked question: Are elements of the scandium and zinc Groups "transition elements"? They are so labeled in the periodic tables of many textbooks of general and inorganic chemistry, yet they haven't the properties usually associated with "transition elements".

The observation that across the *d*-block the subshell of the preponderant type of differentiating electrons is in transition, from being part of valence-shells to being part of atomic cores, at least in the case of zinc and cadmium, holds, also, for the *p*-block, in the case of neon, but not, completely, for the inner-transition elements of the *f*-block.

One scholar of the history of the phrase "transition elements" recommends that it be abandoned: "[I]t is at variance with the older chemical usage and the term *d*-block . . ." (14). In addition, it is at variance (as noted above) with the shape of the Left-Step Periodic Table, as well as (vide infra) the shapes of the Right-Step Periodic Table (Figure 55) and Mendeleev's Secondary Kinship Table (Figure 23).

The Conventional Table has also given rise to the table-specific phrases "pre-transition elements" and "post-transition elements". The LSPT's periods have no "pre-transition elements". What periods of the LSPT have, on proceeding within them left-to-right, as one moves downward in the LSPT to longer and longer periods, are increasing numbers of "transition-elements" prior to occurrence of "non-transition elements".

23. Regularity and "The Periodic System". Any Periodic Table that, like the LSPT, has no irregularities is graphic expression of the fact that the Periodic System is, above all else, a S*YSTEM*, deemed, by Mendeleev, to be, accordingly, *natural,* not arbitrary. (Application of Whewell's definition of "natural" awaited development of atomic physics.) Consideration of the properties of only helium, neon, and beryllium in placement of helium in the Periodic System does not do justice to the System's regularities. Similarly, location of La and Ac beneath Sc and Y (15) greatly distorts the overall shape of all perfectly regular periodic tables. Regularity of tabular expression of the Periodic Law is an instance where truth and beauty coincide, and guard against slides down the slippery slope of irregularities.

Irregularities beget irregularities. With helium above neon, twenty-eight (to sometimes thirty) elements footnoted, and blocks of groups arranged irregularly, by size, it may seem allowable — indeed, even desirable, on chemical grounds — to place hydrogen above carbon! (16).

Many remarks cited herein in support of He/Be hold also for H/Li. The converse, however, is not true. Mendeleev's reason for placing hydrogen in the same family as the alkali metals, because they have the same maximum state of oxidation, does not, of course, hold for helium and the alkaline earth metals. Yet, in that respect helium and those metals, jointly, are part of a number of higher-order trends (**11, 91**) that run from the *s*-block through the *p*- and *d*-blocks to the *f*-block. An apparent defect in the LSPT, in a narrow view, turns out, again, in a broader view, to be a virtue.

24. Different Punctuations of Mendeleev's Line. Some periodic tables place a single *element* in two places: e.g., hydrogen above lithium (owing to its having, like lithium, one valence-shell electron and a maximum oxidation number of +1) and above fluorine [owing to its having, like fluorine, one valence-shell vacancy and a minimum oxidation number of –1 (Appendix VI)]. Other tables have an entire *Group* in two places: usually the Noble Gas Group, on the left, before the alkali metals, labeled Group 0 (owing to the historic nobility of its elements); and on the right, after the halogens, labeled Group 8 (owing to xenon's highest state of oxidation, XeO_4). Sometimes Mendeleev placed the coinage metals, simultaneous, in two different locations. Romanoff's sfdps table of 1934, displayed as our 9th figure below, has an entire *block* in two places: the *s*-block, on the left (so that Group labels, based on highest states of oxidation, start at 1); and on the right (so that blocks' dimensions — widths and heights — change in an orderly manner).

```
                                                                           H  He
 H  He                                                                     Li Be
 Li Be                                                 B  C  N  O  F  Ne|Na Mg
 Na Mg                                                 Al Si P  S  Cl Ar|K  Ca
 K  Ca                         Sc Ti V  Cr Mn Fe Co Ni Cu Zn Ga Ge As Se Br Kr|Rb Sr
 Rb Sr                         Y  Zr Nb Mo Tc Ru Rh Pd Ag Cd In Sn Sb Te I  Xe|Cs Ba
 Cs Ba|La Ce Pr Nd Pm Sm Eu Gd Tb Dy Ho Eu Tm Lu Hf Ta W  Re Os Ir Pt Au Hg Tl Pb Bi Po At Rn|Fr Ra
 Fr Ra|Ac Th Pa U
```

In all periodic tables, the order in which atomic orbitals of different types (*s*, *p*, *d*, or *f*) are occupied with increasing atomic number is -

$$s\ s\ p\ s\ p\ s\ d\ p\ s\ d\ p\ s\ f\ d\ p\ s\ f\ d\ p\ s$$

The orbital sequence may be punctuated in several ways. The two commonest punctuations, illustrated above in Figure 9, start or stop with the *s*-block.

Punctuation by Max. Oxid. No.:	*s sp sp sdp sdp sfdp sfdp*
Punctuation by Blocks' Sizes:	*s s ps ps dps dps fdps fdps*

As discussed in Appendix VII -

> *The s-block plays a distinctive role in the Periodic System.*

It is somewhat "whacky" (Appendix VIII).

Ending periods with *s*-elements yields, directly, the LSPT. Starting periods with *s*-elements requires, for congener verticality, creation of gaps within periods, starting with the second period. Produced is Werner's "long form" version of today's Conventional Periodic Table (Figure 2, bottom left; Figure 9, left rectangle).

Other punctuations of the sequence *sspspsdpsdpsfdpsfdps* are illustrated by the second and third figures on the left in Figure 2. Two sentiments regarding the *fdps* punctuation are cited in Appendix IX.

25. Dyads and Triads. Every period of the Left-Step Periodic Table is paired, by length, with an adjacent period. The paired-periods are called "dyads" shown below (Figure 10), in bold face, in the format of the Conventional Periodic Table, for Groups 1st triads.

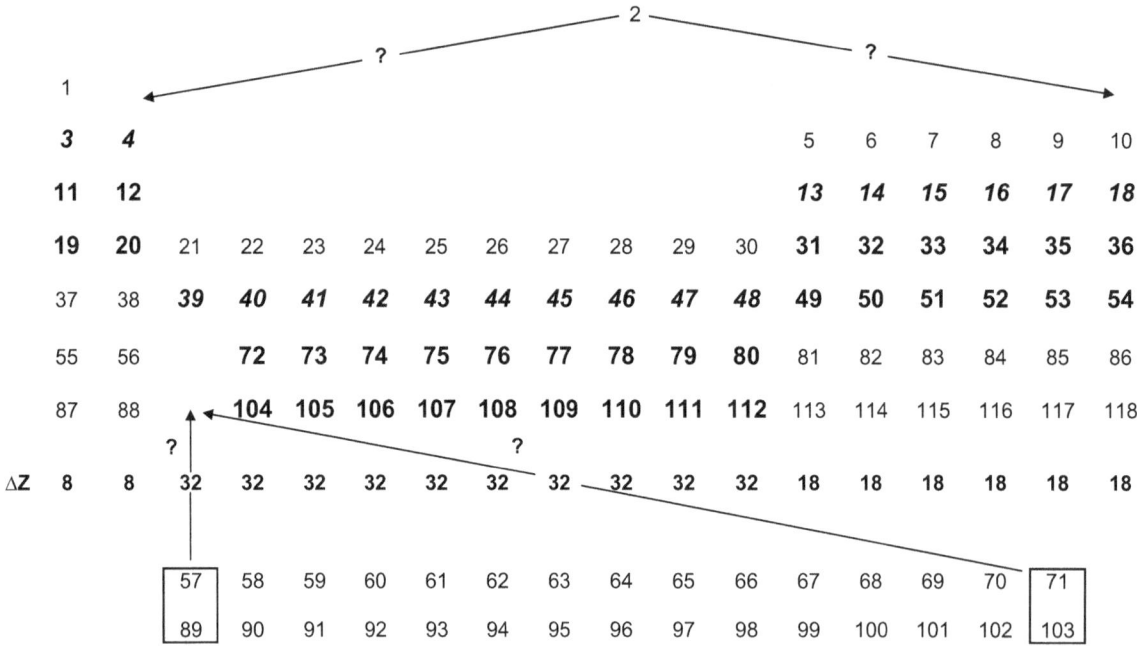

Triads, discovered by Dobereiner in 1816 (17), were the first intimations of chemical periodicity. In modern dress, atomic numbers of triads' members replace atomic weights. Approximate relations among atomic weights become exact relations among the atomic numbers of triads' members:

$$Z_2 - Z_1 = Z_3 - Z_2 \quad \rightarrow \quad Z_2 = (Z_1 + Z_3)/2.$$

That arithmetical exactitude, together with triads' distribution in the Periodic System, resolves in a simple manner the problem of the problem elements, via the simple -

<u>Even-Number Rule</u>

Applicable to all Periodic Tables that have helium above beryllium.

A Group's Triads begin with its 2ⁿᵈ, 4ᵗʰ, and 6ᵗʰ members.

<u>Triad Even-Number Lemma</u>

Groups' first members are not members of conventional triads.

Conventional triads are triads in which all its members belong to the same Group.

Dobereiner noted, e.g., that fluorine does not form a triad with Cl and Br. "[It] indeed belongs to the salt-forming elements," he wrote, "but certainly not to the group of chlorine, bromine, and iodine" (17). Illustrated by Dobereiner's last assertion is the magnitude of first-element distinctiveness in the *p*-block (Sections **30, 31,** and **41**).

Figure 11 below illustrates why dyads yield triads that obey the Even-Number Rule.

Figure 11

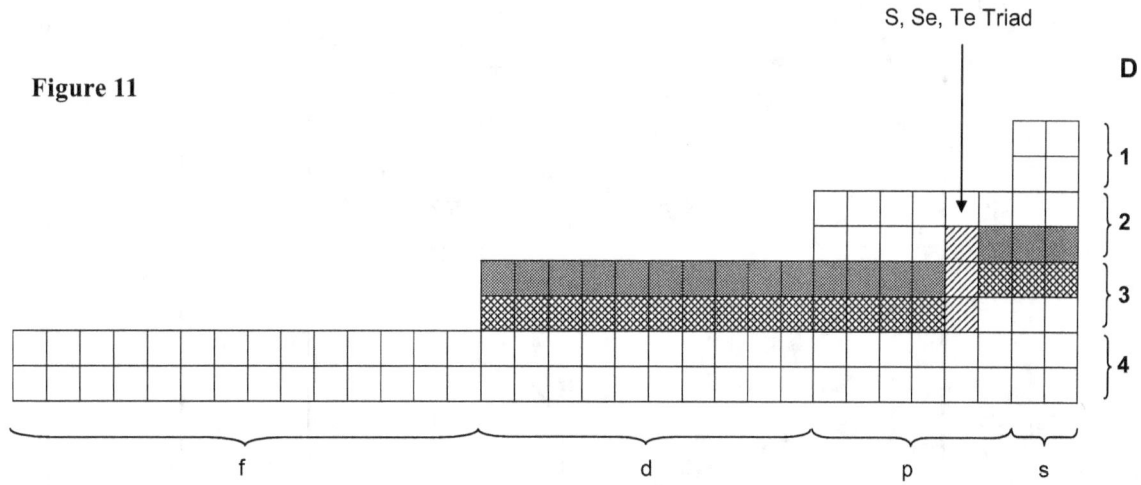

Another consequence of Periodicity's dyadic character is the -

<u>Triad ΔZ Rule</u>

First triads of all of a block's Groups have the same ΔZ-values.

Triads' ΔZ-values exhibit a block-to-block trend:

Block:	*s*	*p*	*d*	*(f)*
ΔZ (first triads):	8	18	32	(50)

If D is the ordinal number of the dyad of a triad's first member, then for that triad -

$$\Delta Z = 2(D + 1)^2$$

The Triad Rules are displayed graphically, for Group I, in Figure 12 below.

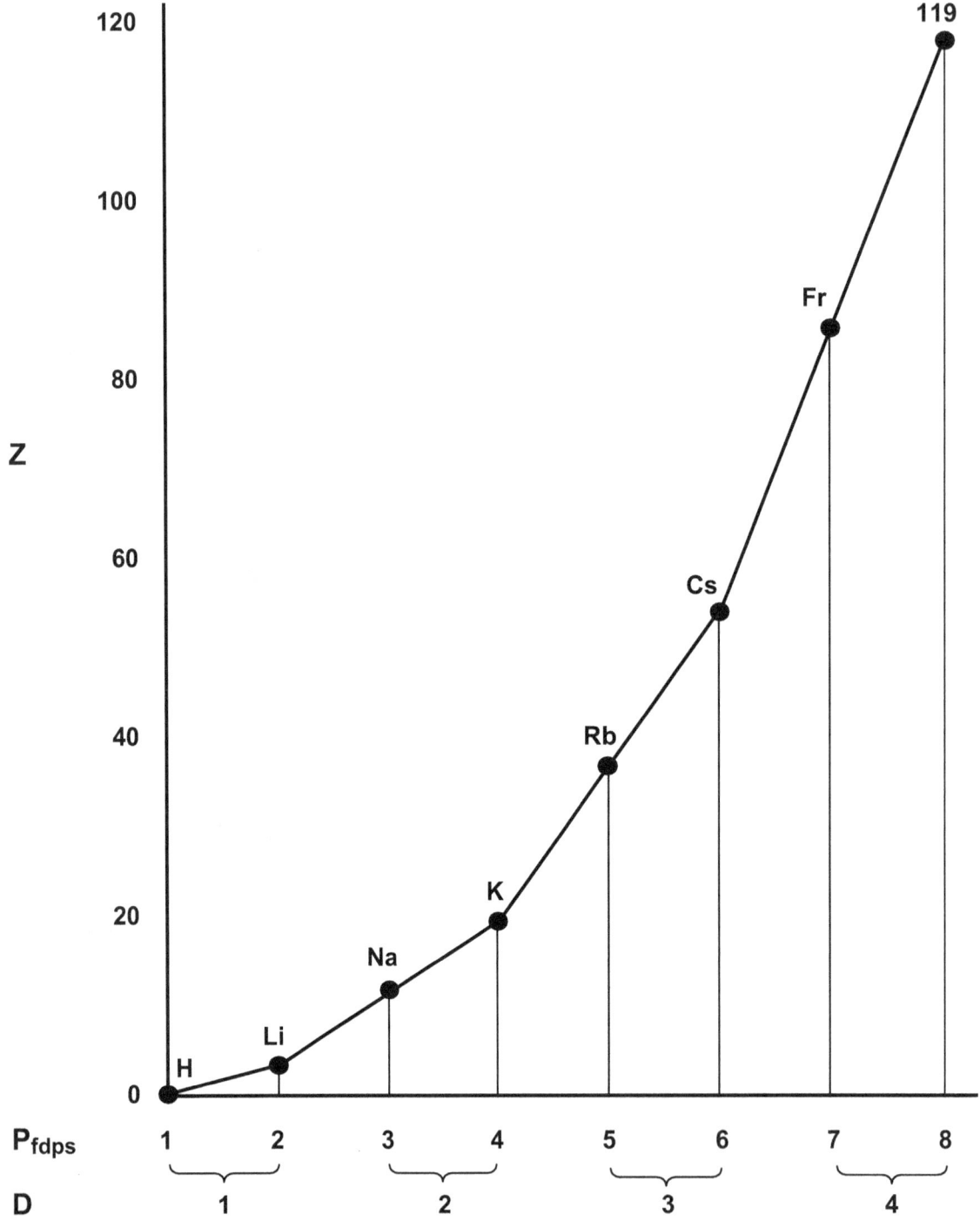

Placement of La and Ac beneath Sc and Y and He above Ne (and H above F) violates the Triad Rules (Figure 10).

Placement of Lu and No beneath Sc and Y and He above Be (and H above Li) satisfies the Triad Rules.

Placement of hydrogen above carbon (16) creates numerous violations of the Triad Rules, in the carbon and lithium Groups. H/C also violates the Rule of Group Sizes (**27**), the LSPT's overall shape, Mendeleev's maximum oxidation state criterion for Group membership, Mendeleev's "absolute distinction" (**34**), a Mendeleev-Jensen rule of first-element distinctiveness (**30, 31**), and a Row-Orbital Correspondence (**48**). From the standpoint of those regularities, hydrogen-above-carbon is as unsatisfactory an assignment as helium-above-neon. All miss-assignments of elements in periodic tables violate essentially the same set of rules.

Bassett proposed a dyadic periodic table in 1892 (18). Except for its location of the alkali and alkaline earth metals, it's essentially a mirror image of the LSPT: a right-step table, read left-to-right, top-down. Unlike Mendeleev's tables of that time, it makes, correctly, thorium, uranium, and assumed trans-uranium elements congeners of the lanthanides (unsuspected by Mendeleev). Bassett omitted hydrogen, perhaps because its inclusion would have destroyed his table's dyadic character. No periodic table that, like Bassett's table, begins periods with the alkali metals (not aligned vertically in Bassett's table, except in pairs, as dyads) exhibits dyadic character at the outset.

26. Physical Basis of the Triad Rules. So simple a resolution, by the Triad Rules, of over a century of debate regarding locations in periodic tables of the problem elements might seem, initially, too simple to be true. Triads are, however, as mentioned, a direct manifestation of Periodicity's dyadic character, expressed algebraically by the Left-Step Table's Equation of Form (**49**):

P(period number) = *r*(row number within a block) + 2ℓ *(block number)*

Chemically, the "2" of the product 2ℓ encodes occurrence of dyads, which yield triads. Physically, the "2" encodes effects on orbital energies of screening by inner-shell electrons of valence-shell electrons of different degrees of penetration, owing to different angular quantum numbers ℓ (**61**). In summary -

Triad Rules reflect fundamental physical phenomena.

The Triad Rules are a simple route to He/Be. An even simpler arithmetical route becomes obvious once one accepts the Principle of Occurrence of Groups in "Blocks" of Widths 2(2ℓ + 1).

27. Groups' Sizes. Placement of helium above beryllium in a modern periodic table yields a simple rule regarding Groups' sizes. For Z_{max} = 120 -

Group Size = 8 - 2ℓ

He/Ne creates two exceptions to the Group Size Rule, one in the *p*-block, for the Ne Group, and one in the *s*-block, for the Be Group (as does, similarly, H/F; and H/C). Location of H and He in a "special block" of their own, floating above periodic tables, creates two exceptions to the Rule in the *s*-block. In a completely regular system, however, there's nothing "special" or exceptional about any part of the system. Therein, however, lies a major weakness, from a chemical point of view, of the irregularity-free LSPT. Without embellishment, such as in Figure 8, or without oxidation-number-based column labels (**94**), the LSPT does not indicate in any way whatsoever the *s*-block's chemically distinctive character, particularly that electrons in outermost *s*-orbitals are valence-shell electrons throughout *f*-, *d*-, and *p*-blocks.

For periodic tables that have helium above neon instead of above beryllium, the Rules of Triads and Group Size hold for all Groups except the neon and beryllium Groups. Using those Rules to locate helium above beryllium is a simple, rational way of using Periodicity's global regularities to solve "The Helium Problem".

28. Classifications of Groups by Size. Because of Periodicity's dyadic character (LSPT step-height equal to 2), movement of helium from its position in the LSPT above beryllium to neon ("for chemical reasons") makes the Neon Group in the $\ell = 1$ block *the same size* as the Beryllium Group in the $\ell = 0$ block (provided the bottom period is complete). Both Groups then have seven members, contrary to the Group Size Rule (**27**). Produced is the notion — when, in addition, the lanthanides and actinides are footnoted — that the Periodic System contains, by size, merely two types of Groups: "Main Groups" and "sub-Groups". It's another linguistic infelicity (along with the phrase "transition metals") identified with the aid of the LSPT. Actually, there are, by size (for Z through 120), four types of groups. Following conventional terminology, one might call them "main-main" Groups, "mm" (the $\ell = 0$, *s*-block Groups); "main" Groups, "m" (the $\ell = 1$, *p*-block Groups); "sub"-Groups, "s" (the $\ell = 2$, *d*-block Groups); and "sub-sub"-Groups, "ss" (the $\ell = 3$, *f*-block Groups). Correspondingly, there is a block-to-block trend in the character of the orbitals of the electrons that differentiate an element from its predecessor in Mendeleev's Line, from "outer-outer", "oo", and "outer", "o", orbitals, to "inner", "i", and "inner-inner", "ii", orbitals, Figure 13, below.

Figure 13.

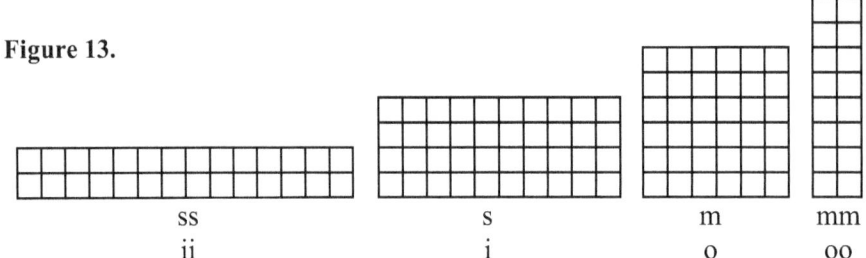

	ss		s		m	mm
	ii		i		o	oo

29. Redundancies and Other Aids to Understanding Periodicity: a Review and a Look Ahead. Understanding, it's said, is seeing that the same thing expressed different ways is the same thing. Understanding the Periodic Classification of the Elements includes seeing that –

- Periodicity's Dyadic Character and its Rules of Triads and Group Sizes are essentially three different ways of saying the same thing: *Chemical Periodicity has a regularity — expressible numerically, algebraically, and graphically — that makes helium a congener of the alkaline earth metals* (**25**).

- Helium-above-beryllium is, logically speaking, the same thing as hydrogen-above-lithium. Hence, if H/Li is logical, then so is He/Be.

- Helium-above-neon is essentially the same thing, logically speaking, as hydrogen-above-fluorine. Hence, if H/F is not acceptable, as a primary kinship, then neither is He/Ne.

As the author's daughter has put it:

> H/Li ➔ He/Be.   Not-H/F ➔ Not-He/Ne.

- Recognition of tertiary chemical kinships (**80**) is the same thing as redefining helium's kinship to neon.

- The leading integers generated by Chemical Periodicity (**45**) have the same mathematical properties as the leading quantum numbers of atomic physics (**48**).

- Madelung's diagram of the order of orbital occupancy in Bohr's Aufbau Process is essentially the same thing as the LSPT (**71**).

- Step-pyramid and "short-form" periodic tables and alphanumeric column labels exist for essentially the same reason (**79, 94**).

- All periodic tables that feature the same Groups are essentially the same thing (**100**).

- Highly relevant for the block order *fdps* is the irrelevance of the metal-nonmetal dichotomy in the Classification of the Chemical Elements *according to the Periodic Law* (**19**).

- Highly relevant for the helium question is Mendeleev's –
    - Distinction between artificial and natural classifications (**20**).
    - Absolute Distinction (**34**).
    - Rule that, although all elements are individuals, some, in upper-most regions of periodic tables, are more so than others (**30**).

30. Mendeleev's Rule of Light-Element Distinctiveness. "[E]lements of small atomic weight are characterized by sharply defined properties," states Mendeleev's *Principles of Chemistry* (**19**), as translated from the 7th Russian edition. One wonders: Was Mendeleev translated correctly? For, of course, *all* element (when pure) have "sharply defined properties". Mendeleev intended to say, it would seem, that –

> *Relative to their congeners*, all elements of small atomic weight have highly *distinctive* properties.

By "small atomic weight" Mendeleev meant elements H through F.   [Fluorine is sometimes called a "super-halogen", as it differs, according to Mendeleev, in many points from the other ("true") halogens, Cl, Br, I, as, e.g., in not having any positive states of oxidation]. Ionization energies and proton affinities of the noble gases illustrate that, from the standpoint of the Rule of First-Element Distinctiveness, the Noble Gas Group's first element is neon, not helium, Figure 14, below.

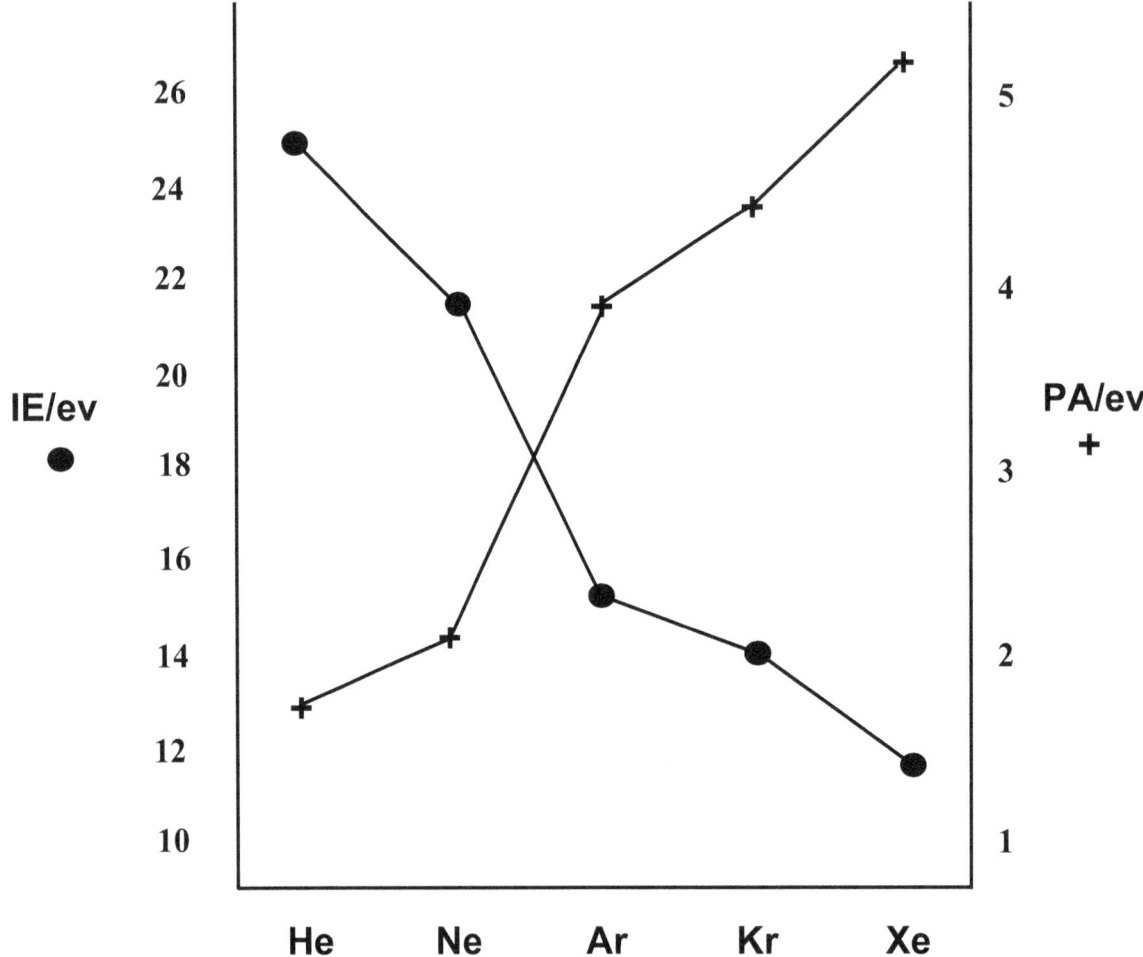

Mendeleev's Rule of Light-Element Distinctiveness, applied to helium, is, arguably, the most important overlooked feature of the chemistry of the elements. One reviewer (2) of an early version of a portion of this report referred to Mendeleev's Rule as "the [present] author's *so-called* 'First-Element Rule'" (emphasis added). (All of Mendeleev's "light elements" are first-row elements in periodic tables.) Ironically, even when the Rule *is* featured in a textbook (1), along with the statement that "It can now be seen that there is a direct and simple correspondence between [atoms'] electronic structure and the form of the periodic table" (1), helium appears, nonetheless, without notice or apology, above neon.

> *With the development of an electronic explanation of periodic tables, helium-above-neon has become, in the light of Mendeleev's Rule of Light-Element Distinctiveness, one of the leading contradictions in the history of chemical thought.*

The most familiar example of Mendeleev's Rule of Light-Element Distinctiveness is Nature's lightest element: hydrogen. Hydrogen's distinctiveness with respect to its congeners is extraordinary, in accordance with –

31. Jensen's Extension of Mendeleev's Rule of Light-Element Distinctiveness (20).

> "[E]lements in the first row of any . . . block tend to show abnormalities relative to elements of later rows of the same block. . . [T]he degree of divergence increases in the order
>
> $$s\text{-block} \gg p\text{-block} > d\text{-block} > f\text{-block}"$$

Consider, e.g., the first elements of each block: H, B, Sc, and La. In their properties hydrogen and lithium diverge more from each other than do boron and aluminum, which diverge more from each other than do scandium and yttrium, which diverge more from each other than do lanthanum and actinium.

Block-to-Block Trend in Distinctiveness of Blocks' First Elements

H (w/r Li) >>> **B** (w/r Al) >> **Sc** (w/r Y) > **La** (w/r) Ac

Within a group, distinctiveness of an element with respect to its next heavier congener generally decreases downward.

Trend in Distinctiveness within Groups of the s-Block

$$H \;\;\ggg\;\; Li \;\;\gg\;\; Na \;\;>\;\; K$$
$$He \;\;\ggg\;\; Be \;\;\gg\;\; Mg \;\;>\;\; Ca$$

Similarly, the p-block exhibits such trends in distinctiveness as -

$$B \gg Al > Ga \quad N \gg P > As \quad O \gg S > Se \quad F \gg Cl > Br.$$

Distinctiveness' two trends — its vertical trend, within Groups (Mendeleev), and that trend's horizontal trend, block-to-block (Jensen) — are displayed in the format of the LSPT in Figure 15 below.

Figure 15.

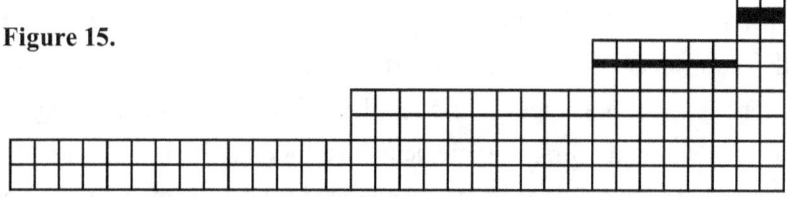

Called to mind is the pattern of hydrogen-atom's energy levels, inverted, Figure 16 (next page).

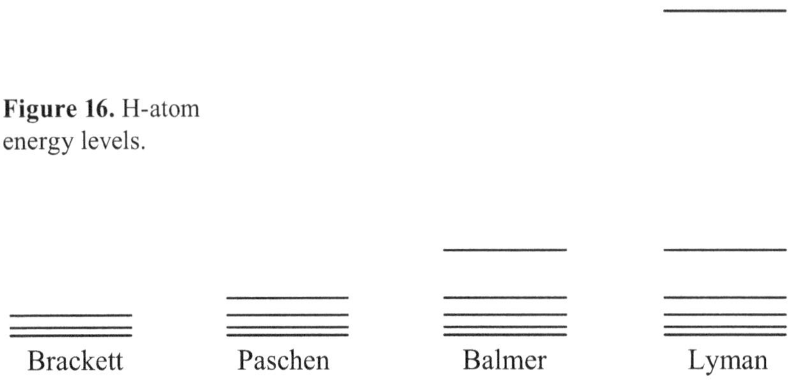

Figure 16. H-atom
energy levels.

| Brackett | Paschen | Balmer | Lyman |

Line spacing in Figure 16 above corresponds to line thickness in Figure 15 (previous page).

The *s*-block ($\ell = 0$) corresponds to the Lyman series (principal quantum number of the lowest level, $n_1 = 1$), the *p*-block ($\ell = 1$) to the Balmer series ($n_1 = 2$), the *d*-block ($\ell = 2$) to the Paschen series ($n_1 = 3$), and the *f*-block ($\ell = 3$) to the Brackett series ($n_1 = 4$). Orbitals of differentiating electrons of blocks' first-row elements are, correspondingly, 1s, 2p, 3d and 4f. Illustrated is a similarity between leading features of the descriptive chemistry of the elements and the Balmer-Rydberg expression:

$$1/n_1{}^2 - 1/n_2{}^2$$

The correspondence between, on the one hand, from descriptive chemistry, trends in elements' distinctiveness and, on the other hand, from atomic spectroscopy, the Balmer-Rydberg expression suggests that the Periodic Law is, in essence, about atoms — in agreement with Mendeleev's views (**34**). Similarly, Bohr's account of the Balmer-Rydberg expression suggests that the "natural interpretation" of "chemical periodicity and the periodic table" lies, as is well-known, today, "in the detailed electronic structure of the atom" (1).

32. Physical Explanation of First-Row Distinctiveness. "The first-row distinctiveness rule has a clear physical origin," writes professor Frank Weinhold (21). "[Differentiating electrons of] [i]naugural elements of an new ℓ-block are able to occupy orbitals with no lower core orbitals of the same symmetry. Those orbitals are therefore free of the powerful Pauli exchange repulsions of inner core orbitals of the same symmetry. While all electrons are subject to the weaker ($1/r$-type) Coulombic core repulsions to electrons of other symmetry, only the first-row pioneers of each ℓ-block are exempted from the requirement to preserve strict orthogonality to inner-shell core orbitals of the same symmetry, the strong (exponential-type) 'steric repulsions' of Pauli exchange origin. In the absence of core exchange repulsions, the first-row elements are therefore able to contract to anomalously small valence orbital radii, with lower energy, enhanced nuclear screening, and other distinctive characteristics compared to second- and subsequent rows. For the *s*-block, the first-row distinctiveness is even stronger, because *neither* Coulombic nor exchange repulsions to prior core orbitals are present. The first-row distinctiveness rule therefore reflects one of the most important general principles of electronic *Aufbau* and chemical periodicity, and its obscuration by misplacement of He in the *p*-block must be counted a significant disadvantage of the [conventional] periodic table."

Numerical indices of elements' distinctiveness *(41)* follow from two general remarks by Mendeleev: one regarding an "absolute distinction"; the other regarding an invariant property of atoms. Immediately below is a generalization of the Mendeleev-Jensen Rule.

33. Distinctiveness of First Occurrence of Periodic Tables' Features. Periodic tables have six kinds of features: over 100 *elements* grouped into 32 *Groups* arranged in 4 *blocks* of 20 *rows* in 8 *periods* (for the Left-Step Table) arranged in 4 *dyads*.

> *Each feature's first occurrence in periodic tables is particularly distinctive.*

The first element in the Periodic System is the only active element whose atoms have no inner-shell electrons. Periodic tables' first row (and first period), H He, contains the two elements that, relative to their congeners, have the most distinctive properties. The first dyad (H & He, Li & Be) is the only dyad whose elements contain only s-type electrons. The first Group (H and the alkali metals) is the only group that has a maximum oxidation number of +1 (and, consequently, is usually the first Group in periodic tables). And the first block (the *s*-block) is Periodicity's most distinctive block (Appendix VII).

In summary, what Mendeleev said of "elements of small atomic weight" holds for all of periodic tables' components that have small ordinal numbers: 0 or 1.

Another leading Principle of Periodicity, emphasized by Mendeleev, yet frequently overlooked, at the present time, is immensely important in placement of helium in periodic tables. Mendeleev stated the principle as a distinction.

34. Mendeleev's "Absolute Distinction". "The central idea that aided me in undertaking the study of the periodic law of the elements," wrote Mendeleev in 1899 (22), "consists primarily in [the] absolute distinction between an *atom* [of, e.g., the element carbon] and a *simple body* [such as diamond or graphite]."

Mendeleev's Pronouncement Regarding the Subject of the Periodic Law

> "All that I am going to say [about the Periodic Law] must be understood as relating to *atoms* . . . not to *simple bodies*." (22)(e).

Earlier Mendeleev had written, in famous footnote 11 of Volume II of *Principles (19):*

"[The Periodic Law] *expresses the properties of real elements* [i.e., atoms], and not of what may be termed their manifestations visually known to us [as 'simple bodies']. The external properties of elements and compounds are in periodic dependence on the atomic weights of the elements only because these external properties are

31

themselves the result of the properties of the real elements, which unite to form the 'free' elements and their compounds."

The Periodic Law, in other words, is about helium, carbon, and oxygen *atom*s, not inert gases, diamonds, and ozone. Indeed, -

The phenomenon of allotropy immediately implies that the Periodic Classification of the Elements is not a classification of the elements as simple substances.

Elsewhere Mendeleev remarks that "it is very important to note [that the Periodic Law is about] the real elements, and not the elements in the free state as we know them." The concept "molecular weight", he adds, is proper to elements in the free state [such as nitrogen, oxygen, the halogens, yellow phosphorus, and rhombic and monoclinic sulfur, whereas the concept] "atomic weight" refers [obviously] to atoms.

Sometimes the hardest things to see, artists say, are the things that stare us in the face. Mendeleev's reference to "atomic weights" in his statement of the Periodic Law implies straight away that the Classification of the Chemical Elements According to the Periodic Law is a classification of *atoms*, not simple substances. The present report concerns, accordingly, chiefly *atomic properties,* such as atomic numbers, atomic refractivities, atomic first-stage ionization energies, atomic energy levels, atomic electronegativities, atomic orbitals, atomic core charges, atomic core radii, atomic electron shells (and subshells), numbers of valence-shell electrons, number of valence-shell vacancies, atomic ratios in elements highest oxides, and quantum mechanical descriptions of atomic structure.

"Your account of periodicity is pure numerology", complains an advocate of placement of all elements in periodic tables according to the Principle of Adjacency for Similar Simple Substances. Exactly! Guilty as charged. Numeration is a natural consequence of atomic theory. "Where is the chemistry in your discussion of periodicity?" asks our critic. In emergence from the test tube of Mendeleev's Line (Figure 1) — an abscissa, so to speak, whose companion ordinate is this report.

You have no exact knowledge [at least in physics] Rutherford famously said, defining "exact knowledge", unless you can express it in numbers. You have no fundamental knowledge of atomic theory, one might add, unless you can express it in integers. Mendeleev's signal achievement was a natural classification of the elements that generates sequences of natural numbers (**45**) that correspond in several instances to leading quantum numbers of atomic physics (**48**).

All of mathematics, it's said, can be based, by logic, on the natural numbers. All one might wish to know about the chemical elements can, in principle, be calculated, by quantum mechanics, from the elements' ordinal numbers in Mendeleev's Line.

35. Inputs from Atomic Physics. Mendeleev's Classification of the Elements according to the Periodic Law was based on two *atomic* properties: *atomic* weights and *atomic* combining capacities; i.e., maximum states of oxidation. Always cited by Mendeleev in his periodic tables in *Principles,* adjacent to or beneath his groups' numerals, are the chemical formulas of chemistry's "seven types" of "Higher Saline Oxides", displayed in his famous Chapter XV ("The Grouping of the Elements and the Periodic Law") in the following fashion [Roman numerals added] (23):

I	II	III	IV	V	VI	VII
Na_2O	Mg_2O_2	Al_2O_3	Si_2O_4	P_2O_5	S_2O_6	Cl_2O_7
	or		or		or	
	MgO		SiO_2		SO_3	

Early periodic tables were called, accordingly, "oxidation tables", with elements grouped according to maximum states of oxidation, regardless of whether or not they were solids or gases, light or heavy, rare or abundant, combustible or noncombustible, metallic or nonmetallic **(19, 20)**.

Mendeleev's atomic-weight/oxidation-state tables had two puzzling features: (i) atomic-weight inversions at Ar & K, Ni & Co, and Te & I; and, contrary to Mendeleev's Rule of Light-Element Distinctiveness, (ii) location of inert, low-boiling helium above inert, low-boiling neon. Moseley's investigations, undertaken to resolve the Ni-Co issue, rescued the Periodic System from the atomic-weight anomalies. Atomic spectroscopy and quantum physics have rescued the System from the He/Ne Light-Element Rule Anomaly. One might suppose that, having welcomed the former rescue, chemists would welcome the latter one. In fact, the He/Ne Anomaly is not only seldom discussed in the chemical literature — as being anomalous from two points of view: one chemical (an exception to Mendeleev's Rule of Light-Element Distinctiveness); one physical (an exception to classification of atoms according to numbers and types of outer electrons). Usually it's not even acknowledged. It's just not. That's "sort of amazing," writes a nonchemist.

36. Atomic Refractivities and Locations of H and He in the Periodic System. Post-Mendeleev developments support his contention that the Periodic Law is a Law of Atoms. Conventional wisdom considers, to repeat, that "Chemical periodicity and the periodic table now find their natural interpretation in the detailed electronic structure of atoms" (1). Measurements of the behavior of atoms' electron clouds in electric fields support Mendeleev's rule regarding first-element distinctiveness, provided H and He are placed at the head of the *s*-block, Figure 17 below.

Atomic Refractivities
(cc/mole)

H	Li	Na	K	Rb		(H)	F	Cl	Br	I
1	13	23	43	66		(1)	1.6	6	8	14
He	Be	Mg	Ca	Sr		(He)	Ne	Ar	Kr	Xe
0.5	5	14	26	33		(0.5)	1	4	6	10

From the standpoint of atomic refractivities and Mendeleev's Rule of First-Element Distinctiveness, -

- o F and Ne *are* satisfactory first-elements in their Groups whereas H and He, at the head of the halogens and noble gases, *are not.*

- o H and He at the head of the alkali and alkaline earth metals are, in their distinctiveness there, better "first elements" than Li and Be.

- o First-element distinctiveness is greater in the *s*-block (for H and He) than in the *p*-block (for F and Ne).

Taken with the Mendeleev-Jensen Rule of First-Element Distinctiveness, atomic refractivities indicate several times over, as do ionization energies and proton affinities (Figure 14), that the natural location for hydrogen and helium in periodic tables is at the top of the *s*-block.

37. An Absolute Prohibition. Mendeleev's Rule of First-Element Distinctiveness (**30**) and his "absolute distinction" (**34**) yield an –

Absolute Prohibition

Two light elements,   such as helium and neon,   similar as simple substances,   but dissimilar atomically,   cannot belong to the same Group.

"Same Group" means, for the most part (see **90**), -

- o same number of valence-shell electrons

- o same type(s) of valence-shell electrons (except for several elements in the *d*- and *f*-block)

- o similarity as simple substances, *except in the case of light element*s

He/Ne violates all three statements. He/Be satisfies them.

The answer to The Helium Question, Which elements are helium's congeners?, is not, according to Mendeleev, Which *simple bodies* is helium like? but, rather, <u>Which *atoms* are helium's atoms like?</u> Left open is the question: Alike in what respect(s)? Number and type(s) of valence shell electrons? *And,* in addition, alike in number of valence-shell vacancies (**24**)? The answer is Yes to the last question, if one seeks an electronic definition of Group membership that corresponds to placement in periodic tables of hydrogen and helium in a block of their own, separate from the alkali and alkaline earth metals. The answer is No, if one accepts the logic of the LSPT. The answer is Yes-and-No, if one wishes to embrace the fact that two-dimensional slices through the multi-dimensional space of

Chemical Periodicity may yield different periodic tables that may correspond to different electronic definitions of Group membership.

Something there is about Periodicity (its multi-dimensionality) that leads to the question: Which is best? "Which" may be: a definition; column labels; a periodic table. The best answer always seems to be (**94, 99**) the same as the answer in mathematics to the question: Which coordinate system is best? Use whatever system or scheme seems most useful *at the moment*.

38. Justification for He/Ne During the Early History of the Periodic System. Because Mendeleev could not answer the question What atoms are helium atoms like?, it was natural for him to place helium in the same family as the other inert gases. Indeed, it was probably necessary to do so in the early days of the Periodic System, in order that [to paraphrase Penrose regarding Bohr's dated views regarding quantum mechanics (25)] the Theory of Similarity of Congeners (Heavy Congeners, that is) could actually be used, and progress in development of the Periodic System could be made. Yet (to continue in a paraphrase of Penrose), as one can see today, such an assignment could only be a temporary one, and does not resolve the question of why, and in what instances, Atomic Physics and the Theory of Congener Similarity do not agree.

39. Unusual Associations in Periodic Tables. At a time (2006) in the history of the Periodic System when, for over a century, helium has been, and usually still is, located in periodic tables above neon, it perhaps cannot be over-emphasized that from the standpoint of the properties of simple bodies and the Rule of Congener Similarity, the Periodic Classification of Atoms creates unusual associations for first-row elements in the *p*-block and, even more so, in the *s*-block. Not alike as simple bodies but, nonetheless, alike atomically, and, therefore, in the same Group in a classification of the elements *based on the Periodic Law of Atoms* (26), are boron and aluminum, carbon and lead, nitrogen and phosphorus, oxygen and sulfur, and, particularly noteworthy regarding unlikeness as simple bodies, hydrogen and lithium, and helium and beryllium.

> *Nature's most striking examples of Mendeleev's Rule of Light-Element Distinctiveness are her two lightest elements.*

Mendeleev's second rule that leads to quantitative indices of elements' distinctiveness concerns a well known property of atoms.

40. "Something Immutable" about Atoms. "Carbon, in the atomic state," wrote Mendeleev (27), "possesses . . . something immutable, whether it takes the form of charcoal or is contained in carbon dioxide, whether it appears as a shiny diamond or is contained in the multitude of organic compounds . . . Carbon in this state is not a concrete body, and yet it is a material substance possessing a certain number of properties."

Today one can say, as Mendeleev could not, that carbon's "immutable" "material substance" is the *nucleus of a carbon atom* plus, to an excellent approximation, throughout most of chemistry, its *inner-shell electrons*. In short -

An atom's "immutable" "material substance" is its <u>core</u>.

A core's leading properties are its charge, type(s) of valence-shell electrons, and size. Core charge is always the same among congeners. Character of valence-shell electrons is usually the same. Always different, within a Group, are core sizes, Figure 18.

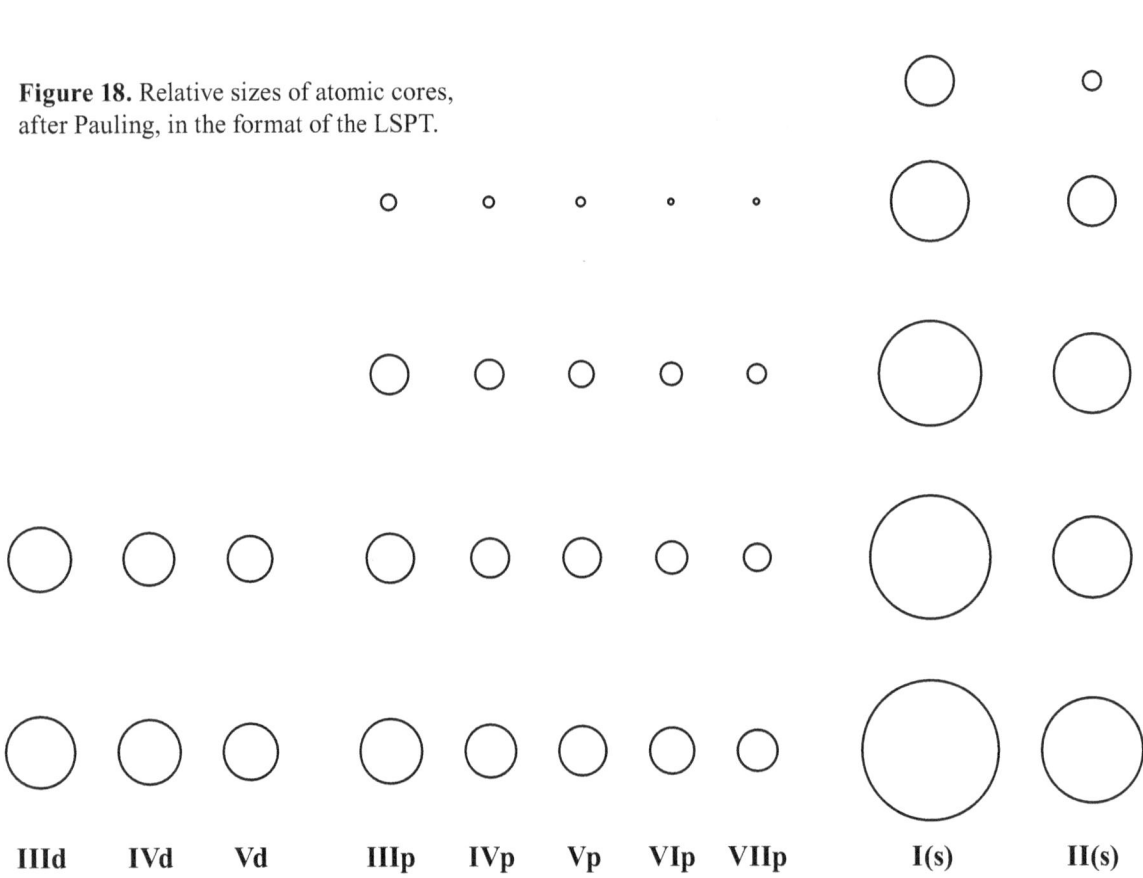

Figure 18. Relative sizes of atomic cores, after Pauling, in the format of the LSPT.

IIId	**IVd**	**Vd**	**IIIp**	**IVp**	**Vp**	**VIp**	**VIIp**	**I(s)**	**II(s)**

41. Numerical Indices of Similarity and Distinctiveness. Immediately apparent in Figure 18 is first-element distinctiveness in the *p*-block and especially in the *s*-block. Ratios of core radii yield Indices of Similarity, S. Fractional changes in radii yield Indices of Distinctiveness, D, Figure 19 (following page).

Ratios of Core Radii

r_1/r_2

Indices of Similarity

					B	C	N	O	F	H	He
					0.40	0.36	0.32	0.31	0.27	0	0
					Al	Si	P	S	Cl	Li	Be
					0.81	0.77	0.72	0.69	0.67	0.63	0.48
Sc	Ti	V	Cr	Mn	Ga	Ge	As	Se	Br	Na	Mg
0.83	0.85	0.84	0.84	0.82	0.76	0.75	0.76	0.75	0.78	0.71	0.66
Y	Zr	Nb	Mo	Tc	In	Sn	Sb	Te	I	K	Ca
0.97	1.00	0.96	0.98	1.00						0.90	0.88
Lu	Hf	Ta	W	Re						Rb	Sr
										0.88	0.84
										Cs	Ba

La	Ce
0.92	0.92
Ac	Th

Fractional Changes in Core Radii

$(r_2 - r_1)/r_2$

Indices of Distinctiveness

					B	C	N	O	F	H	He
					0.60	0.64	0.68	0.69	0.73	1	1
					Al	Si	P	S	Cl	Li	Be
					0.19	0.23	0.28	0.32	0.33	0.37	0.52
Sc	Ti	V	Cr	Mn	Ga	Ge	As	Se	Br	Na	Mg
0.17	0.15	0.16	0.16	0.18	0.24	0.25	0.24	0.25	0.22	0.29	0.34
Y	Zr	Nb	Mo	Tc	In	Sn	Sb	Te	I	K	Ca
0.03	0.00	0.04	0.02	0.00						0.10	0.12
Lu	Hf	Ta	W	Re						Rb	Sr
										0.12	0.16
										Cs	Ba

La	Ce
0.08	0.08
Ac	Th

Cited for the *f*-block in Figure 19 above are S- and D-values for only the first two elements (La and Ce). Cited for the *d*-block are values for the first two elements of its first five Groups. Cited for the *p*-block are values for the first three elements of its first five Groups. Illustrated are the Mendeleev-Jensen Rules of First-Element Distinctiveness (**30, 31**). *Consistent with chemical intuition, D-values generally decrease downward within Groups.*

D-values suggest, weakly, an extension of Jensen's Extension of Mendeleev's Law of Light-Element Distinctiveness. Along a row within a block, D-values generally increase in the same direction as they do from block-to-block: i.e., from left-to-right. Fluorine, for instance, which, unlike chlorine and the other halogens, has no positive states of oxidation, has a D-value of 0.73, larger than the D-value of 0.69 for oxygen, which, like sulfur and the other chalcogens, and unlike fluorine, does have a positive state of oxidation.

Fractional changes in core radii make hydrogen and helium the most distinctive elements in the Periodic System. Core radii do not, however, distinguish between H and He. According to maximum states of oxidation, helium, considered a congener of beryllium, is more distinctive than hydrogen, considered a congener of lithium. Arguably -

> *Helium is the most distinctive element in the Periodic System, when it is located in the s-block above beryllium.*

That statement agrees with the intra-block extension of Jensen's inter-block extension of Mendeleev's intra-Group Law of Elements' Distinctiveness.

Distinctiveness increases upward within Groups, rightward (weakly) within blocks, and, in the LSPT, rightward from block-to-block. Broadly speaking, -

> *In the Left-Step Periodic Table, distinctiveness of an element with respect to its congeners increases from lower left to upper right.*

In displays of trends in congeners' distinctiveness and elements' electronegativities, the *fdps*, Left-Step Periodic Table and the *sfdp*, Long-Form of the Conventional Periodic Table are complementary. In the *fdps* Table, distinctiveness tends to increase diagonally upward, left-to-right, and electronegativities jump around. In *sfdp* Tables, electronegativities tend to increase diagonally upward, left-to-right, and distinctiveness jumps around. Those limitations of the particular tables and their joint complementarity arise from the special character of the *s*-block (Appendix VII), which contains the elements of highest distinctiveness (H and He) and the elements of lowest electronegativities (the alkali metals).

42. Second- and Third-Element Distinctiveness. The fact that beryllium has long been considered to be the most highly distinctive element, with respect to its congeners, in its Group has been cited as ruling out, by the Rule of First-Element Distinctiveness, placement of helium above beryllium. For then one would have *two* highly distinctive elements in the same group! Of course, the same objection should be leveled at placement of hydrogen above lithium, long considered to be noteworthy in its distinctiveness with respect to the other alkali metals. Numerical indices of distinctiveness reveal the error in the objection's logic. *Degrees of distinctiveness exist among congeners beyond the distinctiveness of Groups' lightest elements.*

Numerical indices of distinctiveness (**41**) agree with the distinctiveness, cited above, of the second-row elements lithium and beryllium in the *s*-block. In the *p*-block, high affinity for oxygen sets the second-row elements Al, Si, P, and S apart from their heavier congeners. "[T]he resemblance of antimony to arsenic," wrote Odling (28), "is greater than the resemblance of arsenic to phosphorus. Moreover," continues Odling, in a remark regarding third-element distinctiveness, "the co-resemblances of caesium, rubidium, and potassium, and of barium, strontium, and calcium . . . are far greater than are the co-resemblances of potassium and sodium, and of calcium and magnesium." Regarding sodium, Mendeleev wrote that "Na and K, Li and Rb . . . have been frequently placed among the alkaline metals, although there exists between K and Na . . . more difference of properties than there is between K, Rb, and Cs" (29).

Distinctiveness of blocks' third-row elements — the *s*-block's Na and Mg; the *p*-block's Ga, Ge, As, Se, Br, and Kr; and the *d*-block's Lu, Hf, Ta, W, Re, Os, Ir, Pt, Au, and Hg — arises from Periodicity's dyadic character. It makes blocks' third rows the first rows of their blocks to follow rows in blocks to their left. Atomic cores of elements in the third rows of the *s*-, *p*-, and *d*-blocks are the first ones to be completed by, respectively, p^6, d^{10}, and f^{14} subshells. Produced by Periodicity's dyadic character is the phenomenon of Secondary Periodicity (**95**).

43. Quantification of Chemical Intuition. The greater elements' distinctiveness, the greater, broadly speaking, is their importance to life, and the greater, consequently, is the space devoted to them in chemistry textbooks (except, of course, in the case of helium, which has no chemistry — beyond the statement that it has no chemistry). Mendeleev begins his discussion of the elements in *Principles* with chapters on H, O, N, and C, in that order, D-values 1, 0.69, 0.68, and 0.64, respectively. Listed below is the -

Order of Average D-Values along Blocks' Rows through the First
Ten Types of Atomic Orbitals Occupied in Bohr's Aufbau Process

1s >>> 2p >> 2s > 3s > 3p > 4p > 3d > 4s,5s > 4f > 4d
(1)   (2)   (3)   (4)   (5)   (6)   (7)   (8)   (9)   (10)

Notes:

(1)   The extraordinary instances of H and He.

(2)   The majority of Mendeleev's "light elements". Most of the most widely discussed elements in chemistry textbooks.

(3)   The remainder of Mendeleev's "light elements". The first members of Periodicity's leading examples of "diagonal relationships".

(4)   Periodicity's leading examples of third-element distinctiveness. In *Principles* Mendeleev devotes an entire chapter to sodium.

(5)   Inorganic textbooks often have separate chapters on Si, P, S, and Cl.

(6)   Often discussed in inorganic textbooks with 5p and 6p elements.

(7)   Often singled out for special discussion in inorganic textbooks, before discussion of 4d and 5d elements.

(8)   Usually discussed together in inorganic textbooks.

(9)   Often discussed with 5f elements.

(10)   Usually discussed with 5d elements.

44. Caveat: Limits of Core-Size-Based Indices of Distinctiveness. Indices of distinctiveness based on radii of atomic cores indicate degrees of distinctiveness among elements *of the same Group* but not, as the following core radii show, among elements of different Groups. Quite different from each other are the elements whose cores are paired below, vertically, by charge and size.

Sc^{+3}	Ti^{+4}	V^{+5}	Cr^{+6}	Mn^{+7}
0.81	0.68	0.59	O.52	0.46

In^{+3}	Sn^{+4}	Sb^{+5}	Te^{+6}	I^{+7}
0.81	0.71	0.62	0.56	0.50

Important, along with a core's charge and size, is its instantaneous shape: i.e., its "configuration of maximum probability" (30). The Mn^{+7} core has an outer p^6 subshell, the I^{+7} core an outer d^{10} subshell. Nooks and crannies, so to speak, about a p^6 subshell are vacant in Mn^{+7}, occupied in I^{+7}, by 4d electrons. Accordingly, unshared valence-shell electrons occupy about Mn^{+7} stereochemically inactive, nook-and-cranny-filling, delocalized 3d orbitals. About I^{+7}, unshared valence-shell electrons occupy localized, stereochemically active, conventional coordination sites. Accordingly, the corresponding elements have decidedly different chemistries.

45. Periodicity's Integers. Numerical indices of distinctiveness based on core radii of congeners arise from a union of Group membership (from descriptive chemistry) and partitions of interatomic distances (from x-ray crystallography). Periodicity itself generates several sets of *integers*, Figure 20 (following page).

Listed in Table 1 below are symbols for Periodicity's integers, their chemical significance (Figure 20), and their physical significance [discussed later **(48)**].

Table 1. Chemical and physical significance of integers generated by chemical periodicity.

And I cherish more than anything else the Analogies, my most trustworthy masters. They know all the secrets of Nature, and they ought to be least neglected in Geometry.

KEPLER

Integer of Periodicity	Symbol	Physical Interpretation
Element Ordinal Number	Z	Atomic Number
Block Ordinal Number	ℓ	Angular Quantum Number
Row Ordinal Number within Block	r	Radial Quantum Number
Index of Discontinuity	n	Principal Quantum Number
Ordinal Number of fdps Periods	P	Madelung Parameter $n + \ell$
Dyad Ordinal Number (ON)	D	A Period's largest value of ℓ
Group ON within its Block	e	No. Electrons in Filling Subshell
Row ON within a Periodic Table	R	Orbital ON, Bohr Aufbau Process
Step Height	Δh	Screening Strength Parameter
Row Length, $2(2\ell + 1)$	L_r	Number of Spin Orbitals
Period Length, $2D^2$	L_p	A Periodicity "Magic Number"

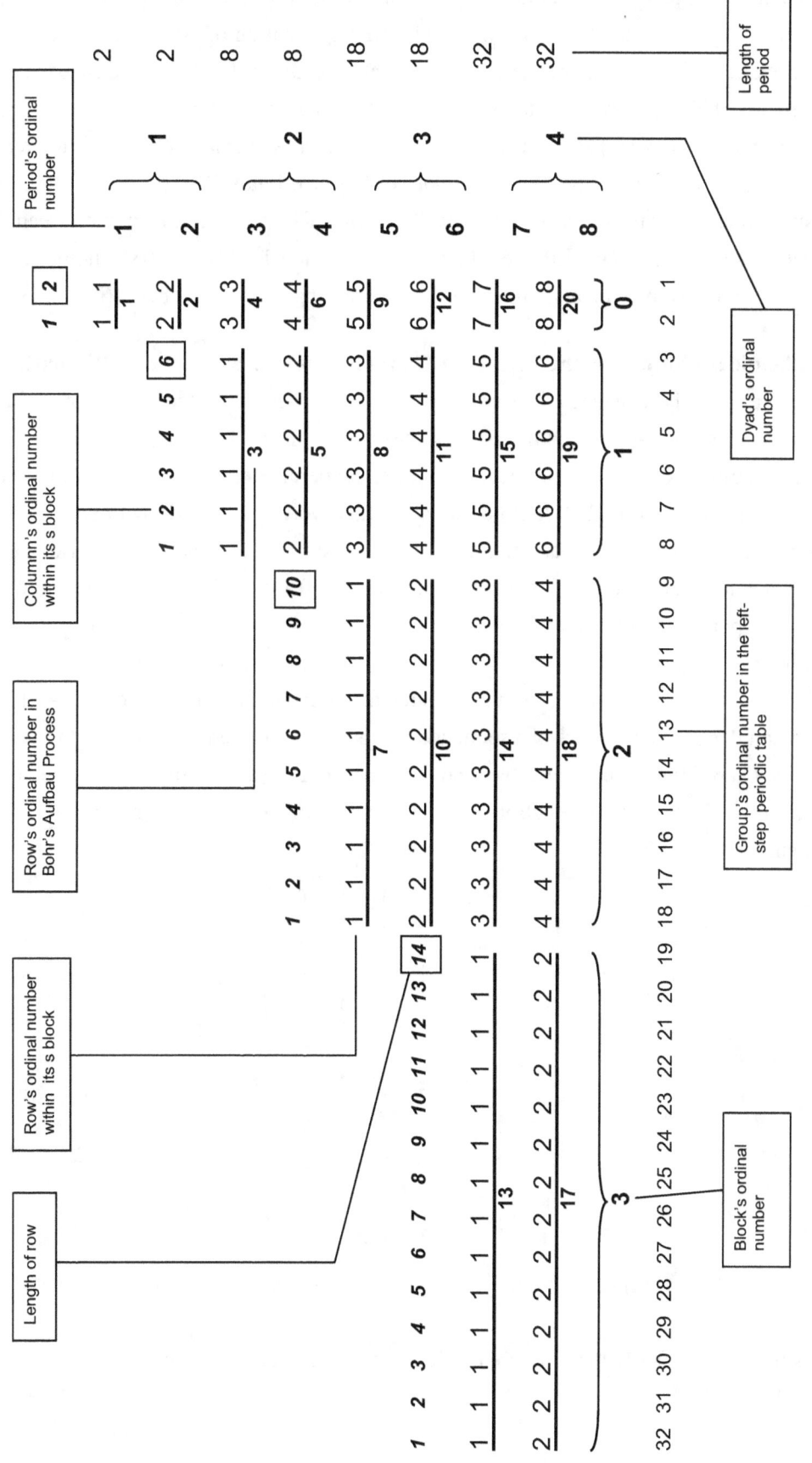

Figure 20. Periocity's Integers.

41

Periodicity's integers are essential for statements of: (i) a Correspondence Principle involving quantum numbers of atomic physics (**48**); (ii) the LSPT's Equation of Form, in various forms (**49**); (iii) a Rule, expressed in various ways, verbally and analytically, of the Order of Occupancy in Bohr's Aufbau Process of blocks' rows and atoms' orbitals (**50**); (iv) graphic expression of the phenomenon of secondary periodicity (**95**); (v) Madelung's Parameter in a form suggestive of a physical interpretation of late occupancy with increasing atomic number of high-ℓ orbitals (**61**); (vi) a modified Madelung Parameter for analysis of the order of atoms' excited states (**69**); (vii) a connection between the LSPT and Madelung's Diagram (**71**); (viii) a set of natural column labels (**94**); and (ix) a numerical index of leading discontinuities in elements properties with increasing atomic number (**46**).

46. Chemical Capture of the Integer *n* from Creation of an Index of Discontinuity. The question arises: Do the discontinuities in elements' properties that occur with increasing atomic number on exiting the *p*-block for the *s*-block correspond in the Left-Step Periodic Table, where those discontinuities occur *within* periods, to a numerical parameter that is solely a function of those integers descriptive of the LSTP? Is it possible, in other words, to encode with some combination of Periodicity's integers the dramatic nonmetal/metal discontinuities that occur periodically in the Periodic System and that are the chief reason for a leading feature of most Periodic Tables: namely, termination of periods at the end of the *p*-block.

Sought is an index that (i) adopts a new value on an entrance to the *s*-block, but that, at the same time, (ii) does not change on an exit from the *s*-block for — eventually, or immediately — the *p*-block and the remainder of the members of a sequence of "main group" elements that are in the same *sfdp* period of the Conventional Periodic Table (and that have, looking ahead, the same principal quantum number for their differentiating electrons). Blocks' row numbers, *r*, satisfy requirement (i) but not requirement (ii), Figure 21.

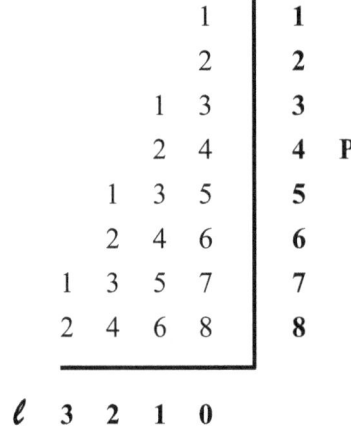

Figure 21. Blocks' rows' *r*-values in the format of the LSPT.

The sum $r + \ell$ satisfies both requirements. It's represented, for reasons explained later (**48**), by the symbol *n*, Figure 22 (next page).

```
                    1  |  1
                    2  |  2
                 2  3  |  3
                 3  4  |  4   P
              3  4  5  |  5
              4  5  6  |  6
           4  5  6  7  |  7
           5  6  7  8  |  8
```

ℓ **3   2   1   0**

Figure 22. Numerical Values of the Index of Discontinuity n in the Format of the LSPT.

On each entrance to the s-block with increasing atomic number, whether at the outset, from the s-block itself, or, as is usually the case, from the p-block, — for each entrance to the s-block $\Delta n = +1$. On the other hand, on exiting the s-block for the p-block, $\Delta n = 0$, whether that passage is immediate ($n = 2$ to $n = 2$ and $n = 3$ to $n = 3$) or delayed, owing to passages through the d-block ($n = 4$ to $n = 4$ and $n = 5$ to $n = 5$) or to passages through both the f- and d-blocks ($n = 6$ to $n = 6$ and $n = 7$ to $n = 7$).

Since $r \geq 1$, the relation $r + \ell = n$ yields the expression $\ell \leq n - 1$. As one sees from inspection of Figure 22, a block's ordinal number ℓ is 1 less than the first (and smallest) n-value cited in its column. Often the "radial quantum number" is defined slightly differently than here (36).

47. Groups' Ordinal Numbers in the LSPT. Not mentioned in Table 1, inasmuch as it has no physical significance, but exhibited at the bottom of Figure 20, is a Group's ordinal number G in the Left-Step Periodic Table. G starts at 1 for the Be Group and ends, at the present time, at 32 for the La Group (31). G is related to a Group's ordinal number, e, in block ℓ by the expression

$$G = 2\ell^2 + 4\ell + 3 - e$$

For the iron Group, e.g., $G = 2(2^2) + 4(2) + 3 - 6 = 13$. As e increases, ℓ constant, G decreases (or course) by the same amount. The change in G for unit increase in ℓ, e constant, is given by the expression –

$$\Delta G = G(\ell + 1, e) - G(\ell, e) = 4\ell + 6$$

ΔG-values for $\ell = 1$ and 2 are 10 and 14. Those are the distances along the horizontal axis of the LSPT between Groups whose elements are related by secondary chemical kinships (**79**).

G-values have modest chemical and historical significance. They track, broadly speaking, Indices of Similarity, **S,** for first-row elements (Figure 19). The larger G, the larger, by and large, is **S.** Also, broadly speaking, if with notable exceptions (notably the Noble Gases), G-values track dates of recognition of Groups. The larger G, the later, in general, recognition of the Group. At the beginning of the G-scale is the $G = 1$, Alkaline-Earth-Metals Group, home of Dobereiner's first triad. At the other end of the G-scale is the La-Ac, $G = 32$ Group, long deemed part of super-Group IIId, or a sub-super-Group IIIf. The lanthanide-actinide pairs of the $G = 19 - 32$ Groups have been the last elements granted Group status in the Periodic System.

48. Physical Significance of Periodicity's Integers r, ℓ, and n. The relations cited above, together with the second factor in the expression $2(2\ell + 1)$ for the number of groups in block ℓ, yield five algebraic forms:

$$\ell \geq 0$$
$$r \geq 1$$
$$n = r + \ell \geq 1$$
$$\ell \leq n - 1$$
$$2\ell + 1$$

Not surprisingly, perhaps, in view of Mendeleev's statement that the Periodic Law is about *atoms*, the algebraic forms cited above, generated phenomenologically from the facts of descriptive chemistry (as summarized in the Periodic Law and exhibited in the LSPT), call to mind quantum numbers of atomic physics, Table 2.

Table 2. Leading Features of Atomic Orbitals' Leading Quantum Numbers.

1. r (radial quantum number) = number of radial nodal surfaces, counting the one at infinity. Hence $r \geq 1$.

2. ℓ (angular quantum number) = number of nucleus-containing nodal surfaces. Hence $\ell \geq 0$.

3. n (total quantum number) = total number of nodal surfaces. Hence, by 1 and 2, $n = r + \ell$. And, hence, by 1 and 3 -

4. $\ell \leq n - 1$. Also -

5. The number of orbitals that have an angular quantum number ℓ is, for any value of n, $2\ell + 1$ (**53**).

Generated by Chemical Periodicity and Atomic Physics is -

A Periodic-Table/Atomic-Orbital Correspondence Principle

> block ordinal numbers ℓ = orbitals' number of angular nodal surfaces
> row ordinal numbers r = orbitals' number of radial nodal surfaces

The orbitals in question are those of a blocks' preponderant type of differentiating electrons.

The first part of the Correspondence Principle is well-known. The second part — hitherto unstated, if not unknown (32) — has been largely overlooked because, in fact, -

> *With helium above neon in periodic tables, the Row-Orbital Correspondence does not hold for either the Noble Gases or the Alkaline Earth Metals.*

Only if helium appears above beryllium in periodic tables do the three integers ℓ, r, and n generated by Chemical Periodicity correspond, *in all instances,* to integers generated by atomic physics.

49. Equation of Form of the LSPT. A leading feature of the LSPT is the simplicity of the algebraic expression of the relation of the ordinal numbers of its periods, P, to those of its blocks, ℓ, and its blocks' rows, r. One sees from Figure 21 that in the $\ell = 0$ block, $P = r$. For $\ell = 1$, $P = r + 2$. For $\ell = 2$, $P = r + 4$. And so forth. For the $\ell = \ell$ block:

$$P_{fdps} \,(= P_{LSPT}) = r + 2\ell = n + \ell$$

The boxed expression immediately above is one of science's centrally productive "magic" formulas (Appendix XI). It expresses in a nutshell the correspondence between the chemical significance of the Integers of Chemical Periodicity and the Quantum Numbers of Atomic Physics. Periodicity's row and block ordinal numbers, r and ℓ, and its Index of Chemical Discontinuity, n, correspond, respectively, to atomic physics' radial, angular, and principal quantum numbers (Table 1). Periodicity's period numbers in the LSPT, P_{LSPT} ($= P_{fdsp}$), correspond to the parameter $n + \ell$ of the Madelung rule for the order of occupancy of orbitals in Bohr's Aufbau Process.

The expression $P_{fdps} = r + 2\ell$ renders obsolete a remark made by Mendeleev in 1899: "The reason for the absence of an explanation concerning the nature of the periodic law resides entirely in the fact that *not a single rigorous, abstract expression of the law has been discovered*" (22, p221, emphasis added). The Bohr-Bury atomic model of 1921 was, as the noun announces, "merely" a model, a "natural interpretation" (1) of observations and not, in a deep sense, an explanation of them. The "rigorous, abstract expression" of periodic law expressed in the form $P_{fdps} = r + 2\ell$ provides, however, a foundation for a qualitative, physical explanation of the shapes of periodic tables (**61**). (The conventional expression $P_{fdps} = n + \ell$ is not as suggestive.)

50. Row-Occupancy Rule Restated. The relation $P = n + \ell$ and the shape of the LSPT yield a restatement of Row-Occupancy Rule (**5**), as a Row-*Orbital* Occupancy Rule.

Row-Orbital Order of Occupation with Increasing Z

> Smallest $n+\ell$ first and for given $n+\ell$ largest ℓ first.

51. The ℓ-Convention. Four reasons may be given, now, for beginning blocks' ordinal numbers at 0 rather than 1, Table 3.

Table 3. Reasons for starting blocks' ordinal numbers ℓ at 0. That choice simplifies statement of -

- an index of discontinuity, n, in elements' properties with increasing atomic number, as $n = r + \ell$ (**46**).

- a correspondence between block's ordinal numbers and atomic orbitals' angular quantum number, as $\ell_{block} = \ell_{orbital}$ (**48**).

- the LSPT's equation of form: $P = r + 2\ell$ (**49**).

- the order of occupancy of atomic orbitals in Bohr's Aufbau Process, as smallest $n + \ell$ first (**50**).

52. Late Recognition of Periodicity's Integers. Mendeleev displayed in his favorite periodic table two sets of integers, Figure 23 (following page).

Mendeleev did not distinguish s-block Main Groups from p-block Main Groups. [Following in his footsteps, neither the American nor the European Group-labeling schemes (Figure 41) distinguish between those two blocks of Groups, although p-block/s-block differences are greater than d-block/s-block differences.] Addition, to Mendeeev's 12 "series", of the number 7, for the s-block's rows, and an additional addition of 2, for rows of the f-block (also unanticipated by Mendeleev), yields a total of $12 + 7 + 2 = 21$ "series", close to the number of rows within blocks of modern periodic tables (for Z through 120), namely: $8 + 6 + 4 + 2 = 20$.

Recognition of a third, and arguably the most obvious, set of integers generated by the Periodic Law came later. In the words of Paneth (33): "[I]mplicit in any table of the periodic system, because here — in contradistinction to its representation by curves (f) — the elements were always arranged in equal distances from each other and not according to actual values of atomic weights" was a consequence stated "with admirable clarity" by the Swedish physicist Rydberg, who, in 1887, wrote that "In investigations on the periodic system not the atomic weights, but *the ordinal numbers of the elements* [emphasis added] are to be used as independent variables."

One of the first investigators to extract a set of ordinal numbers for the elements from the facts of descriptive chemistry was Newlands, in 1865, in his "Law of Octaves". It was "only after 1913, however, when the Rutherford-Bohr theory of the atom had been developed," adds Paneth, that "the deep physical significance of these ordinal numbers, now usually called atomic numbers, could be understood."

It was only after introduction of Janet's Left-Step Periodic Table of 1927, with helium above beryllium, and introduction of Schrodinger's theory of the hydrogen atom, that the deep physical

PERIODIC SYSTEM OF THE ELEMENTS IN GROUPS AND SERIES

	0	I	II	III	IV	V	VI	VII	VIII
1		H							
2	He	Li	Be	B	C	N	O	F	
3	Ne	Na	Mg	Al	Si	P	S	Cl	
4	Ar	K	Ca	Sc	Ti	V	Cr	Mn	Fe Co Ni (Cu)
5		Cu	Zn	Ga	Ge	As	Se	Br	
6	Kr	Rb	Sr	Y	Zr	Nb	Mo	—	Ru Rh Pd (Ag)
7		Ag	Cd	In	Sn	Sb	Te	I	
8	Xe	Cs	Ba	La	Ce	—	—	—	— — — —
9		—	—	—	—	—	—	—	
10	—	—	—	Tb	—	Ta	W	—	Os Ir Pt (Au)
11		Au	Hg	Tl	Pb	Bi	—	—	
12	—	—	Ra	—	Th	—	U		

HIGHER SALINE OXIDES

R	R_2O	RO	R_2O_3	RO_2	R_2O_5	RO_3	R_2O_7	RO_4

HIGHER GASEOUS HYDROGEN COMPOUNDS

				RH_4	RH_3	RH_2	RH	

Figure 23. Mendeleev's favorite ("short form") Periodic Table. A secondary kinship (**79**) table. Integers at the left are ordinal numbers of Mendeleev's horizontal "series". They correspond, broadly speaking, to the ordinal numbers of rows of elements within modern periodic tables' blocks. Roman numerals label Mendeleev's "Groups" — our "Families of Groups" (Jensen). Group VIII contains Mendeleev's ("true") transition elements (**22**).

significance of ordinal numbers of *blocks' rows* — the modern version of Mendeleev's "series" — could be understood.

53. Vector Models of Periodicity's Term (2ℓ + 1). The classical quantum mechanical model of the term (2ℓ + 1), generated by Chemical Periodicity, as half the width of block ℓ, assigns to an electron of angular momentum ℓ, by an assumption of spatial quantization, the 2ℓ + 1 values of the projection of a vector whose index ℓ runs through the integers from ℓ to − ℓ: ℓ . . . 0 . . . −ℓ. An equivalent vector model, according to the newer wave mechanics, associates with the normal to a nucleus-containing nodal surface of an orbital three components: hence existence of 3, one-nodal-surface, ℓ = 1, linearly independent *p*-orbitals. Description of the direction of a second nucleus-containing nodal surface *with respect to the first one* requires 2 additional independent components: hence existence of 3 + 2 = 5 linearly-independent, 2-nodal-surface, ℓ = 2, *d*-orbitals. And so forth. The number of orbitals of type ℓ is, accordingly, 3 + 2(ℓ − 1) = 2ℓ + 1.

54. Chemical and Physical Significance of Periods' Ordinal Numbers. In the LSPT a period number P_{fdps} is the sum of the number of nodal surfaces, $n = r + \ell$, of an orbital of a differentiating electron, with angular surfaces *double-weighted:* $P_{fdps} = r + 2\ell$. The factor "2" in the LSPT's equation of form reflects the Table's dyadic character. It is the key to a physical explanation of Chemical Periodicity (**61**).

For the Conventional, *sfdp* Periodic Table, a period's ordinal number, P_{sfdp}, is equal in the ℓ = 0 block to block row-number r, which is equal to the outer-subshell's principal quantum number n (= r when ℓ = 0), which is also equal to the n-value of the outer-subshell in the period's *p*-block. That is to say -

$$P_{sfdp} = \text{ordinal number of s-block entrance}$$
$$= r\ (\ell = 0)$$
$$= n_{\text{outer subshell in a period's s- and p-blocks}}$$
$$= n_{\text{outer subshells of main group elements}}$$
$$= \text{number of occupied shells in all blocks}$$

55. Relation between P_{fdps} and P_{sfdp}. Transformation of the *fdps* Table to an *sfdp* Table involves, as noted, movement of the *s*-block to the left. That translation leaves P-numbers in the *s*-block unchanged, since the *s*-block is where counting of periods begins, in all periodic tables. To complete the transformation, the *s*-block must be moved down one period, in order to maintain the ordinal order of the elements. (Each "carriage return" in reading a periodic table, left-to-right, *top-down*, involves a movement downward of one period.) That downward movement of the *s*-block decreases by 1 ordinal numbers of periods through the *p*-, *d*-, and *f*-blocks. As one sees from Figure 24 (following page) -

$$\ell = 0:\ P_{sfdp} = P_{fdps}. \qquad \ell > 0:\ P_{sfdp} = P_{fdps} - 1$$

sfdp Table   **fdps Table**

P_{sfdp}	ℓ=0	ℓ=3	ℓ=2	ℓ=1
1	1			
2	2			2
3	3			3
4	4		3	4
5	5		4	5
6	6	4	5	6
7	7	5	6	7

ℓ=3	ℓ=2	ℓ=1	ℓ=0	P_{fdps}
			1	1
			2	2
		2	3	3
		3	4	4
	3	4	5	5
	4	5	6	6
4	5	6	7	7
5	6	7	8	8

Figure 24. *n*-values in the format of the sfdp and fdps periodic tables.

What may appear to be a mere formality regarding P_{sfdp} and P_{fdps} has, in fact, chemical significance. In plots of congeners' properties vs. P for different Groups of a block, it does not matter whether one uses P_{sfdp} or P_{fdps}. When, however, plots for Groups of the *s*- and *p*-blocks are exhibited *together*, regularities, known as Secondary Periodicity, (32) and (**95**), appear throughout the plots when the P-coordinate is P_{fdps} but not when it is P_{sfdp}. That's because P_{sfdp} corresponds to periodic tables in which Periodicity's dyadic character, which is the cause of Secondary Periodicity, is absent at the tables' tops and is elsewhere disjointed, as regards, on the one hand, the *s*-block and, on the other hand, the *p*-, *d*-, and *f*-blocks.

Positive (+) and negative (-) features of *sfdp* and *fdps* periods, sometimes called, respectively, "chemical" and "physical" periods, are listed in Appendix XII.

56. Relation between Period Numbers P and the Principle Quantum Number *n*. The Left-Step Table's equation of form $P_{fdps} = r + 2\ell = n + \ell$ yields -

$$n = P_{fdps} - \ell.$$

As one sees from the left-hand-side of Figure 24, the corresponding expression for the Conventional, *sfdp* Table is -

$$\ell \leq 1: n = P_{sfdp} \qquad \ell \geq 2: n = P_{sfdp} - \ell + 1$$

Along a given period of the *fdps* Table, the larger ℓ the smaller *n*, the smaller ℓ the larger *n*. Starting, for instance, at the left end of period 7 of the LSPT with orbital 4f (always in Period 7, $n + \ell = 7$) and advancing rightward through diminishing values of ℓ, orbitals occur (as shown in Figure 24, on the right) 4f, 5d, 6p, and 7s.

57. Chemical and Physical Significance of Dyads' Ordinal Numbers. Chemically, dyad number D (= 1, 2, 3, or 4, for Z through 120) is the number of blocks in a dyadic periodic table that a period of dyad D traverses. Dyad 3, e.g., traverses 3 blocks: from left-to-right, the *d-*, *p-*, and *s-*blocks. Physically, D is the number of types of subshells, *s*, *p*, *d*, and *f*, occupied in atoms of the dyad.

Each new dyad introduces with its first period a new block and a new type of subshell. Dyad 1 contains only and all atoms that contain only *s* electrons. Atoms of Dyad 2 have for the first time *p* electrons. Together, Dyads 1 and 2 contain all of Mendeleev's typical or representative elements. Dyad 3 contains the first *d-*block ("transition") metals. Dyad 4 contains the first *f-*block ("inner transition") metals. Dyads' ordinal numbers yield a simple expression for lengths of periods in the Periodic System (cf. 34):

$$Period\ Length = 2D^2$$

58. Chemical Significance of Square Pyramid Numbers. Sums of the squares of the integers, starting at 1, are the square pyramid numbers: 1, 5, 14, 30. Four times those numbers — 4, 20, 56, 120 — are the atomic numbers of elements at the ends of dyads: Be, Ca, Ba, 120. That's because the length of a period of dyad D is $2D^2$. Therefore, the ordinal number of an element at the end of Dyad D is -

$$2\sum 2D^2 = 4\sum D^2 = 4[D(D + 1)(2D + 1)/6]$$

Illustrated, once again, in the chemical significance of the square pyramid numbers, is Periodicity's intrinsic regularity and the appropriateness, accordingly, of the Convention of Maximum Regularity in the formative Construction Conventions cited in connection with Figure 1.

To create the LSPT from the expression for the square pyramid numbers, $(2/3)[D(D + 1)(2D + 1)]$, substitute for D the numbers 1, 2, 3, 4. To the resulting sequence, 4 20 56 120, add, by the Triad Rules **(25)**, the intermediate atomic numbers: $(4+20)/2 = 12$, $(20+56)/2 = 38$, and $(56+120)/2 = 88$. Then proceed as in Section **3**.

59. Atomic Numbers of Dyads' Rows' First Elements and Periodicity's 2's. Since a block's ordinal number ℓ is related to its first dyad's ordinal number D_1 by the relation $\ell = D_1 - 1$, and since a block's length is equal to $2(2\ell + 1)$, atomic numbers of blocks' first-row elements start at

$$4[D_1(D_1 + 1)(2D_1 + 1)/6] + 1$$

and end $2(2D_1 + 1) - 1$ elements later. D_1-values of $0 - 4$ yield Z-values of 1 (H), 5 (B), 21 (Sc), 57 (La) and 121. One-half square roots of differences are 1, 2, 3, 4. Suggested is another set of construction rules for the LSPT. Set down the integers 1, 2, 3, 4. Double them. Square the doubles. Then use the squares of their doubles as successive ΔZ-values, starting with Z = 1, Figure 25 (top), next page.

	1	2	3	4	
	2	4	6	8	
ΔZ	4	16	36	64	
Z	**1**	**5**	**21**	**57**	**121**

	1	2	3	4
	1	4	9	16
ΔZ	2	8	18	32
Z_1	1	5	21	57
Z_2	**3**	**13**	**39**	**89**

Figure 25. Generation of atomic numbers of the two elements in dyads' first columns.

Ionization energies of blocks', and dyads', first elements follow a Jensen-type block-to-block trend:

H 13.6   >>>   **B** 8.30   >>   **Sc** 6.54   >   **La** 5.58

To obtain atomic numbers of the first elements of dyads' *second* periods, set down the integers 1, 2, 3, 4. Square them. Double the squares. Then add the doubles of the squares to the atomic numbers of the dyads' first elements, Figure 25 (bottom).

Ionization energies of first elements of blocks' second rows follow another block-to-block trend:

Li 5.39   <   **Al** 5.99   <   **Y** 6.38   <   **Ac** 6.9

Periods of the LSPT's paired-periods start, in summary, with atomic numbers

1 3   5 13   21 39   57 89

The sequence is a superposition of two series. Increments in atomic numbers in each series are based on the integers 1 2 3 4: doubled and squared in one instance; squared and doubled in the other instance.

Squaring-and-doubling and doubling-and-squaring play important roles in the chemical history of Periodic System Systematics. Also, significant features of the physical and mathematical histories of Chemical Periodicity's regularities are expressed in a nutshell by the integer 2. It is a coefficient in the expression $(2\ell + 1)$, owing to orbitals' degeneracies (**53**); a factor in the expression for blocks' widths $2(2\ell + 1)$, owing to the phenomenon of "spin", and the Pauli Exclusion Principle; a coefficient in Madelung's parameter $r + 2\ell$, owing to effects of screening and penetration on orbital energies of many-electron atoms (**61**); the magnitude of step-height of the Left-Step Periodic Table; and it is an exponent in the dyad-length expression $2D^2$, owing to the mathematical fact that sums of the squares of the odd numbers, starting at unity, are the square numbers. It is, also, the atomic number of the element whose location in periodic tables is, at this time, the most controversial and whose presence in its natural location anchors many of Periodicity's regularities.

60. Odd-Row Distinctiveness. Dyads' first periods are the LSPT's odd periods. Those periods contain the first rows of the first blocks through which a Dyad's periods pass, left-to-right. Most distinctive are those first rows. Distinctive, also, are block's third rows, and less so, their fifth rows. Produced, as mentioned, is the phenomenon of Secondary Periodicity (**95**). Blocks' third rows are their first rows that are occupied following occupancy of a new type of subshell. In textbook-discussions of effects of such subshells on subsequently occupied subshells, the order of emphasis is generally:

$$f_{\text{lanthanide contraction}} > d_{\text{scandinide contraction}} > p_{\text{boronide constraction}}$$

61. Physical Explanation for Late Occupancy of *d*- and *f*-Orbitals. Late occupancy of high-ℓ orbitals is a dominant feature of Periodic Tables. It occurs because, unlike energies of orbitals of one-electron atoms, which depend only on n, not ℓ, energies of orbitals of many-electron atoms increase in the order

$$ns < np < nd < nf$$

That order arises, it's generally supposed, because the larger the number of nucleus-containing orbital nodal surfaces, namely ℓ, the less an orbital's electron density near the nucleus, the less it penetrates in many-electron atoms inner-shell electron density, the more its electron density is shielded by inner-shell electron density from the atom's positive nuclear charge, and, hence, the higher the orbital's energy, the later its occupancy in Bohr's Aufbau Process, and, correspondingly, the larger the period number P for the corresponding row in a Periodic Table (35).

62. Power of a Nodal Point at the Origin. A second reason for late occupancy of high-ℓ orbitals in many-electron atoms arises from the fact that, in addition to having r radial and ℓ angular nodal *surface*s, the latter of which pass through atomic nuclei, hydrogen-like orbitals have in the radial part of their wave functions a factor r^{ℓ}, which vanishes at the nucleus for $\ell > 0$, and is small, nearby, particularly for large ℓ. Thus, unlike increases in r, ℓ constant, increases in ℓ, r constant, increase both the number of nucleus-containing nodal *surface*s and, through the orbital's R-equation, the power of the orbital's nodal *point* (r^{ℓ}) at the nucleus.

63. Node-Counting and Madelung's Parameter. The symbolic sum of the three nodal features yields the form of the well-known Madelung parameter $n + \ell$.

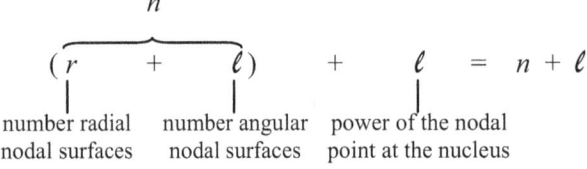

Of course, the sum is, in a numerical sense, purely a formalism. For, as pointed out above, radial and nucleus-containing angular nodal surfaces aren't expected to have the same effects on orbital energies in many-electron atoms. Also, presumably energy-raising effects of the nodal *point* at the nucleus that arise from the factor r^ℓ are different from energy-raising effects of nodal *surfaces*. In the absence of exact knowledge of the numerical contributions that the three terms make to orbital energies, one is limited to saying that the sum of the contributions to orbital energies in many-electron atoms of the two ℓ-terms through unit increases in ℓ is greater than that for unit increases in r. A change in ℓ means, always, two changes: a change in number of nucleus-containing angular nodal surfaces and a change in the power of the nodal point at the nucleus.

Additionally, as will be discussed more fully later **(69)**, the coefficient 1 of ℓ in the Madelung parameter $n + \ell$ for the order of orbital occupancy in Bohr's Aufbau Process is larger than necessary. For instance, for 3d to follow 4s, it is only necessary that ℓ's coefficient k in the expression $n + k\ell$ be greater than 1/2. For 4f to follow 6s, k need only be greater than 2/3.

Perhaps the formal sum of three terms, r and the two ℓ's, one for nodal surfaces, the other for a nodal point, may be viewed as a useful mnemonic regarding orbital features that influence orbital energies in many-electron atoms, particularly the double reason for the importance of the angular quantum number ℓ.

64. Sturm-Liouville/Hartree-Fock Rationale for Node-Counting (and He/Be). "The Madelung rule has a clear mathematical basis in the general Sturm-Liouville theory of one-dimensional wave equations . . . (including the Schrodinger-type equations of chemical interest)," writes Professor Frank Weinhold (21). *Eigenvalues for Sturm-Liouville equations increase monotonically with number of nodes.* "Of course, the 1-dimensional Sturm-Liouville ordering is not immediately applicable to the N-electron Schrodinger equation for atoms," notes Weinhold. However, "the Hartree-Fock approximation allows the complex many-electron Schrodinger equation to be replaced by a one-electron ('independent particle') eigenvalue equation . . . [which] therefore contains the requisite physics to assign a definite electronic configuration to each atom, consistent with the *Aufbau*-type picture that underlies the periodic table" — and placement of helium above beryllium. Briefly put:

Hartree-Fock Approximation + Sturm-Liouville Theory
→ Node Counting → Madelung $r + 2\ell$ Rule
→ Left-Step Periodic Table → He/Be

The usual chemical argument for placement of helium above neon in periodic tables is, in a similar format, -

Helium and neon are inert gases
+
Similarity-Adjacency Rule
→
He/Ne

The argument's logic is impeccable. Its soundness is no better, however, than that of its second premise: that elements vertically adjacent to each other in periodic tables are always similar to each other as simple substances. As is well-known, however, the Similarity-Adjacency Rule is strongly violated by the first-row elements of the *p*-block and, even more so, by hydrogen in the *s*-block.

65. Two Physical Interpretations of the Quantum Numbers *ℓ* and *n*. One might view a nodal *point* at the nucleus as a *degenerate radial nodal surface*, radius zero, and the quantum number *ℓ* as, in addition to the number of nucleus-containing nodal surfaces, the number of degenerate radial nodal surfaces at the nucleus. The principal quantum number *n*, equal to *ℓ + r*, becomes in that interpretation the total number of radial nodal "surfaces", degenerate (*ℓ* in number) plus nondegenerate (*r* in number). The quantum number *n* has, consequently, two physical interpretations. It's the total number of nondegenerate nodal *surfaces* (counting the one at infinity). And it's the total number of radial nodal *modes*, with allowance for the power of the nodal point at the nucleus, Table 4.

Table 4. Chemical and physical significance of the integer *n*.

Chemical

- Index of Discontinuity along Periods of the LSPT

- Sum of blocks' and blocks' rows' ordinal numbers: *ℓ + r*

Physical

- Atoms' Principal Quantum Number

- Sum of an orbital's radial and angular quantum numbers: *r + ℓ*

- Total number of an orbital's nodal *surfaces*

- Total number of an orbital's radial nodal *modes*

- Sole determinant of orbital energies in hydrogen-like atoms

- Determinant with *ℓ* of the order of orbital occupancy in many-electron atoms, via the Madelung parameter *n + ℓ*

66. Dependence of Orbital Energies on *n*, *ℓ*, and *r*. For a hydrogen atom an increase in *n*, *ℓ* constant, increases orbital energy, whereas an increase in *ℓ* or *r*, *n* constant does not. As one sees from the *r/ℓ* spreadsheet of Figure 26 (next page), for the somewhat-hydrogen-like, few-electron atoms He and Li, an increase in *ℓ*, *n* constant, increases orbital energy, but not nearly as much as does an increase in *n*, *ℓ* constant.

r/ℓ	0	1	2	3
1	**1s** 0	**2p** 169081 14903	**3d** 186095 31283	**4f** 191446 36630
2	**2s** 159850 0	**3p** 185558 30925 16956	**4d** 191439 36623 34549 27397	**5f** 193916 39104 37058 30620
3	**3s** 183231 27206 0	**4p** 191211 36470 30267 12985	**5d** 193911 39095 37037 30185	
4	**4s** 190292 35012 25740 0	**5p** 193795 39016 35040 24701	**6d** 195254 40437 38387 31695	
5	**5s** 193341 38300 33200 21027	**6p** 195187 40390 37296 28999		
6	**6s** He 194930 Li 39987 Na 36372 K 27450			

	Z-1	ΔE ≈ 0	Δℓ	Δr	-Δr/Δℓ
He	1	6s --> 6p	+1	-1	1.0
Li	2	6s --> 5f	+3	-4	1.3
Na	10	6s --> 5d	+2	-3	1.5
K	18	6s --> 4d	+2	-4	2.0

He 1s 2s 2p 3s 3p 3d 4s 4p 4d **4f** 5s 5p 5d 5f [**6s** 6p] 6d 6f 7s

Li   2s 2p 3s 3p 3d 4s 4p 4d **4f** 5s 5p 5d [5f **6s**] 6p 6d 7s

Na   3s 3p 4s 3d 4p 5s 4d **4f** 5p [**6s** 5d] 5f 5g 6p 7s 6d

PT 1s 2s 2p 3s 3p 4s 3d 4p 5s 4d 5p **6s** 4f 5d 6p 7s 5f 6d 7p

K   4s 4p 5s 3d 5p [4d **6s**] **4f** 6p 5d 7s 5f 5g 7p

Rb   5s 5p 4d **6s** 6p 5d 7s **4f** 7p 6d 5f 8s

Cs   **6s** 6p 5d 7s 7p 6d 8s **4f** 8p

For Na the *ℓ*-dependence of orbital energy is nearly as large as the *n*-dependence. For K it is larger.

The larger the number of electrons: (i) the larger the effect of *ℓ* on orbital energy E compared to that of *r*; (ii) the later occupancy of high-*ℓ* orbitals, such as **4f**; and (iii) the earlier the occupancy of low-*ℓ* orbitals, such as **6s**. The line labeled "**PT**" gives the order of orbital occupancy with increasing Z in Bohr's Aufbau Process. Lines of constant P_{fdps} (= $r + 2ℓ$) have a slope in Figure 26 of –2. The

Figure's r vs. ℓ layout is one of three such figures. A second one, n vs. ℓ, is the Madelung Diagram (**71**). The third one, P_{fdps} vs. ℓ, is the LSPT.

For hydrogen an increase in ℓ, r constant has the same effect on orbital energy as has an increase in r, ℓ constant. That is to say, -

$$\text{For H:} \quad -(\Delta r/\Delta\ell)_{E\text{ constant}} = 1$$

For helium, the quotient $-(\Delta r/\Delta\ell)_E$ is slightly greater than 1. For other atoms it's significantly greater than 1, the more so the larger the number of electrons. For potassium, the quotient is approximately 2: i.e., $\Delta\ell = 1$ has about the same effect on orbital energy E as $\Delta r = 2$ *(large box, Figure 26)*.

A unit increase in r adds a non-nucleus containing nodal sphere to an orbital whereas a unit increase in ℓ adds not only a nucleus-containing nodal surface but, also, in addition, increases the power of a nucleus-containing nodal point, r^ℓ.

67. Physical Information from Exceptions to the Madelung Orbital-Occupancy Rule in the *d*- and *f*-blocks. Unlike elements of the s- and p-blocks, whose electron configurations follow the Madelung Rule, a number of elements of the *d*- and *f*-blocks exhibit well-known exceptions to the Rule, attributed, in part, to half- and full-subshell effects. The majority of the exceptions, however, do not involve such subshell occupancies. They arise, it's generally assumed, from approximately equal orbital energies in the *d*-block for valence-shell $(n-1)$d and ns orbitals (r equal to, respectively, n-3 and n) and in the *f*-block for valence-shell $(n-1)$f and nd orbitals (r equal to, respectively, n-4 and n-2).

In passing between nearly energy-degenerate $(n-1)$d and ns orbitals, the ratio $-\Delta r/\Delta\ell$ is 3/2. In passing between nearly energy-degenerate $(n-1)$f and nd orbitals, $-\Delta r/\Delta\ell = 2/1$. Both ratios are, as expected, greater than 1, the more so, generally, the greater the number of electrons in the many-electron atoms. Numerical values cited in this section for the coefficient $-(\Delta r/\Delta\ell)_{E \approx \text{const.}}$ from exceptional electron configurations in the *d*- and *f*-blocks are approximately the same as those derived in the previous section from energies of excited states of Na and K.

68. Hypothetical Periodic Tables. Examining consequences of varying the coefficient of ℓ in the Left-Step Periodic Table's equation of form, $P = r + 2\ell$, places the Table in mathematical and physical perspective. The larger hypothetical non-penetration and screening effects are, the larger should be the coefficient of ℓ in the expression (h = column height, Δh = step height)

$$P = r + (\Delta h)\,\ell$$

Thus, along a period ($\Delta P = 0$), if $\Delta\ell = -1$ (as in proceeding left-to-right in the Left-Step Periodic Table), $\Delta r = \Delta h$ (= 2 for the LSPT).

Figure 27 (next page) exhibits periodic tables for $\Delta h = 1$, 2, and 3.

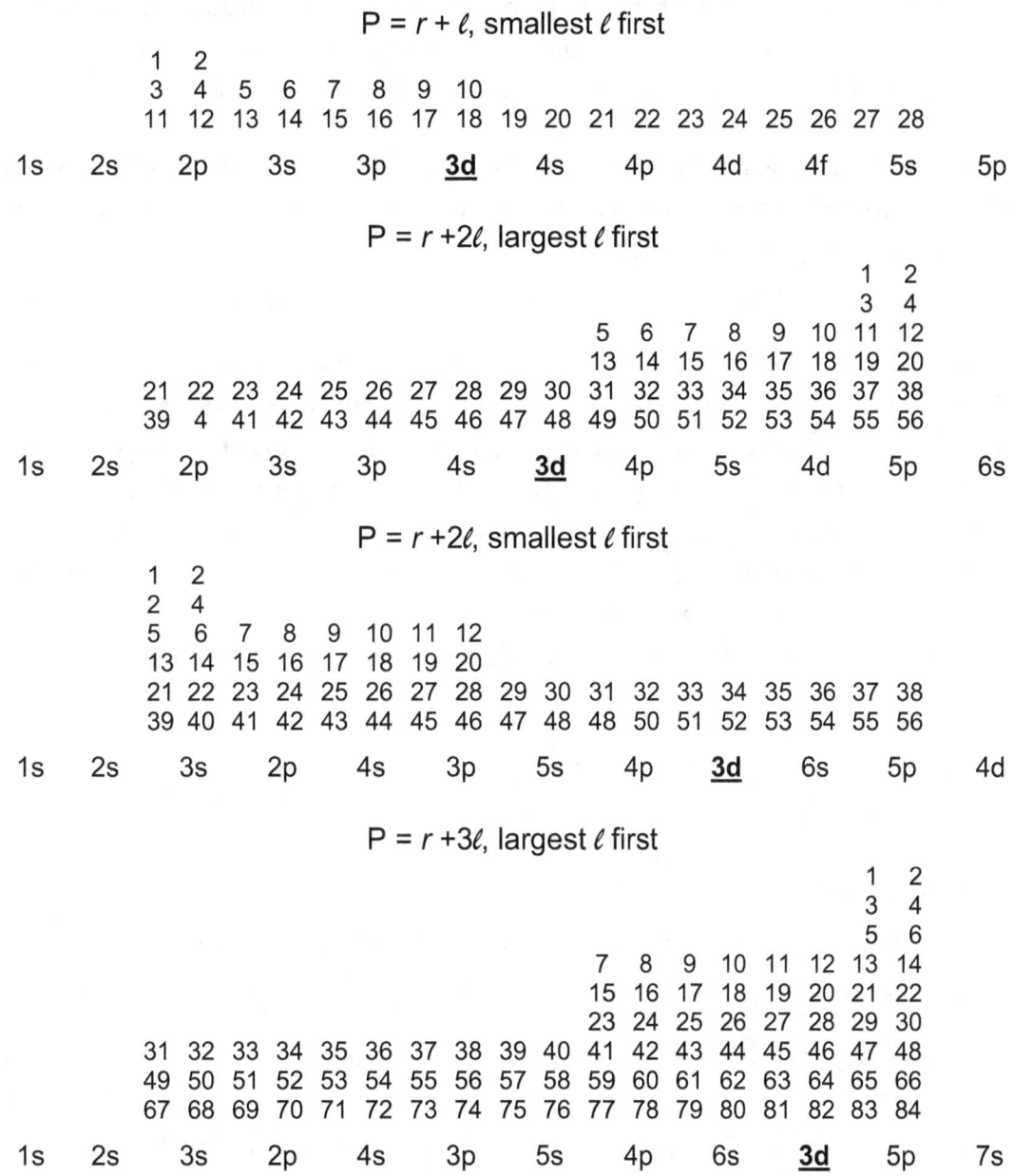

Figure 27. Several hypothetical step-type periodic tables.

Nature's choice, the second table, corresponds, as noted previously, to occurrence of dyads and, within groups, triads, starting with members 2, 4, 6, 8. $\Delta h = 1$ (the top table) yields monadic periods, hence no triads. In the bottom two tables, *first* row elements are members of triads. The bottom table contains tetrads, starting with group members 3, 6, 9, etc.

The larger Δh, the later occupancy of high-ℓ orbitals. One sees in Figure 27 above that for $\Delta h = 1$, smallest ℓ first (top table), 3d precedes 4s and 4f precedes 5s. For $\Delta h = 2$, largest ℓ first (second table), 3d follows 4s and 4f follows 6s. For $\Delta h = 2$, smallest ℓ first (third table), 3d follows 5s and 4f follows 7s. And for $\Delta h = 3$, largest ℓ first (bottom table), 3d follows 6s and 4f follows 8s.

According to the second table in Figure 27, the ground state electron configuration of, e.g., element 8, is $1s^2\, 2s^2\, 2p^4$, whereas, according to the third table, it is $1s^2\, 2s^2\, 3s^2\, 2p^2$. From the point of view of each table, element 8's ground state configuration given by the other table is an excited state.

69. Generalized Madelung Parameter and the Order of Excited States of Atoms and Ions. Results in Sections **67** and **68** suggest comparing orders of orbital occupancy calculated from a generalized Madelung parameter of the form

$$n + k\,\ell = r + (1 + k)\,\ell \qquad k \neq \text{an integer, necessarily}$$

with observed orders of energy levels of atoms and ions, Figure 28 (following page).

Values for k of 0.1, 0.9, and 1.1 generate, respectively, the first three periodic tables of Figure 27, without recourse to the phrase "largest (or smallest) ℓ first". Tie-lines highlight delayed occupancy of nd and nf orbitals with (i) increasing values of k and (ii) increasing numbers of electrons — and, correspondingly, increasingly early occupancy of ns-orbitals. Underlined are hydrogen-like sequences. They tend to increase in frequency and extent in calculated sequences with decreasing values of k and in observed sequences with decreasing numbers of electrons.

Values of k for the calculated $n\ell$ sequences at the top of Figure 28 bracket the following critical values:

$n\ell$ before $n(\ell+1)$ → $k > 0$ ⎫
$n\ell$ before $(n+1)s$ → $k < 1/\ell$ ⎬   Hydrogen-like sequences

ns before $(n-\ell)d$ → $k > 1/2$ ⎫
ns before $(n-2)f$ → $k > 2/3$ ⎬   Periodicity's sequences
$n\ell$ before $(n+1)(\ell-1)$ → $k < 1$ ⎭

$k = 0.9$ reproduces Periodicity's sequence. $k = 1.1$ yields the orbital sequence of the third periodic table of Figure 27.

As k increases, high-ℓ, nf orbitals retreat in the sequences and low-ℓ, ns orbitals advance toward beginnings of sequences.

Listed beneath the calculated sequences are experimental sequences for the alkali metals and several ions isoelectronic with them. Li I (like He I) is completely hydrogen-like. Na I is less hydrogen-like than Li I, K I less so than Na I.

The larger the number of electrons, the larger k needs to be. Na I (10 core electrons) follows the calculated $n+0.4\ell$ sequence for its first 8 excited states. The cation Ca II (18 core electrons) follows the $n+0.6\ell$ sequence through its 10th excited state. Ca I's excited triplet states follow the $n+0.9\ell$ sequence through the 13th excited state.

With decreasing values of k, calculated sequences become, of course, increasingly hydrogen-like. Isoelectronic systems become increasingly hydrogen-like with increasing nuclear charge and, consequently, decreasing relative importance of electron-electron repulsions. Al III is completely hydrogen-like.

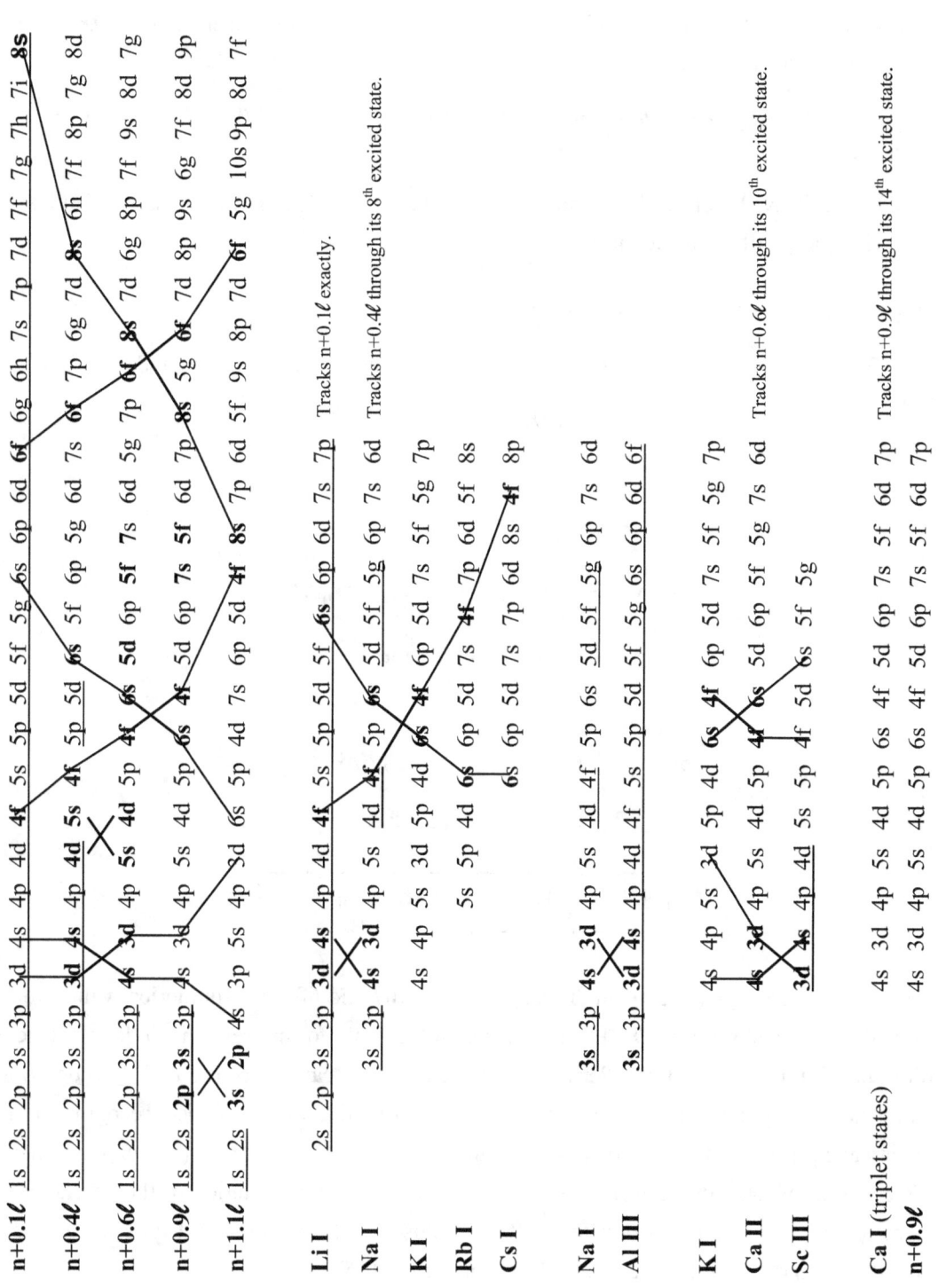

Figure 28. Comparison with experiment of orbital orders generated by (n + kℓ) for different values of k.

In general, however, no single value of k parameterizes an atom's entire manifold of excited states. The sequence of excited states for, for example, K I of 5p 4d 6s 4f 6p 5d 7s, $n+\ell = 6\ 6\ 6\ 7\ 7\ 7$ 7, requires, respectively, k > 1, k < 1, k > 2/3, k > 1, and k < 1.

Similarly, no single value of k is necessary for all parts of periodic tables. Any value greater than zero yields the row-orbital order through element 20, 0.51 the order through element 57, 0.67 the order through element 120. That is to say:

> *The coefficient of ℓ of 1 in the Madelung parameter n + ℓ is, at the outset of Bohr's Aufbau Process, larger than necessary.*

It seems, unlikely, therefore, that an analytical proof of the Madelung Rule, based on fundamental physical principles, will be forthcoming.

70. Relation of k of $(n + k\ell)$ to $-(\Delta r/\Delta \ell)_E$.

Table 5. Assumptions and logic behind a relation between k of $n + k\ell$ and $-(\Delta r/\Delta \ell)_E$.

$$E = E(r, \ell)$$

$$\approx c(n + k\ell)$$

$$= c[r + (1 + k)\ell]$$

$$\rightarrow \quad \Delta E \approx (\Delta E/\Delta r)_\ell \Delta r + (\Delta E/\Delta \ell)_r \Delta \ell$$

$$\approx c\Delta r + c(1 + k)\Delta \ell$$

$$= 0$$

$$\rightarrow \quad (\Delta r/\Delta \ell)_E = - (\Delta E/\Delta \ell)_r/(\Delta E/\Delta r)_\ell$$

$$\approx - c(1 + k)/c$$

$$\boxed{\text{k (of } n + k\ell) \approx - (\Delta r/\Delta \ell)_{E \approx \text{const.}} - 1}$$

Numerical values of $-(\Delta r/\Delta \ell)_E$ in the range 1.5 – 2 (**67** and **68**) yield numerical values for k in the range 0.5 – 1, as observed (**69**). Leading steps on the route to that result include: placement of helium in beryllium's Group in the Periodic System; chemical capture of the LSPT; recognition of integers generated by the LSPT; formulation of a Row-Orbital Correspondence; recognition of the mathematical equivalence to each other of Madelung's diagram and the LSPT (**71**); expression of the LSPT's equation of form in the form P = $r + 2\ell$ (**49**); physical interpretation of the coefficient "2" (**63**); creation of hypothetical periodic tables (**68**); and use of a Madelung-Janet-type spread sheet to examine the dependence of atomic energy levels on r and ℓ (**66**).

71. Diagrams of Janet and Madelung.
Atomic orbitals' r- and ℓ-values are displayed graphically by means of schematic, circle-diameter diagrams in Figure 29 (next page).

○
1011

Figure 29. Orbital diagrams in the format of the LSPT.
   Circles: radial nodal surfaces
   Diameters: nucleus-containing
      angular nodal surfaces

◎
2022

Addresses: $r \ \ell \ n \ $ P
   $n = r + \ell$
   P $= \ell + n$

1123

◎
3033

2134

◎
4044

1235

3145

5055

2246

4156

6066

1347

3257

5167

7077

2358

4268

6178

8088

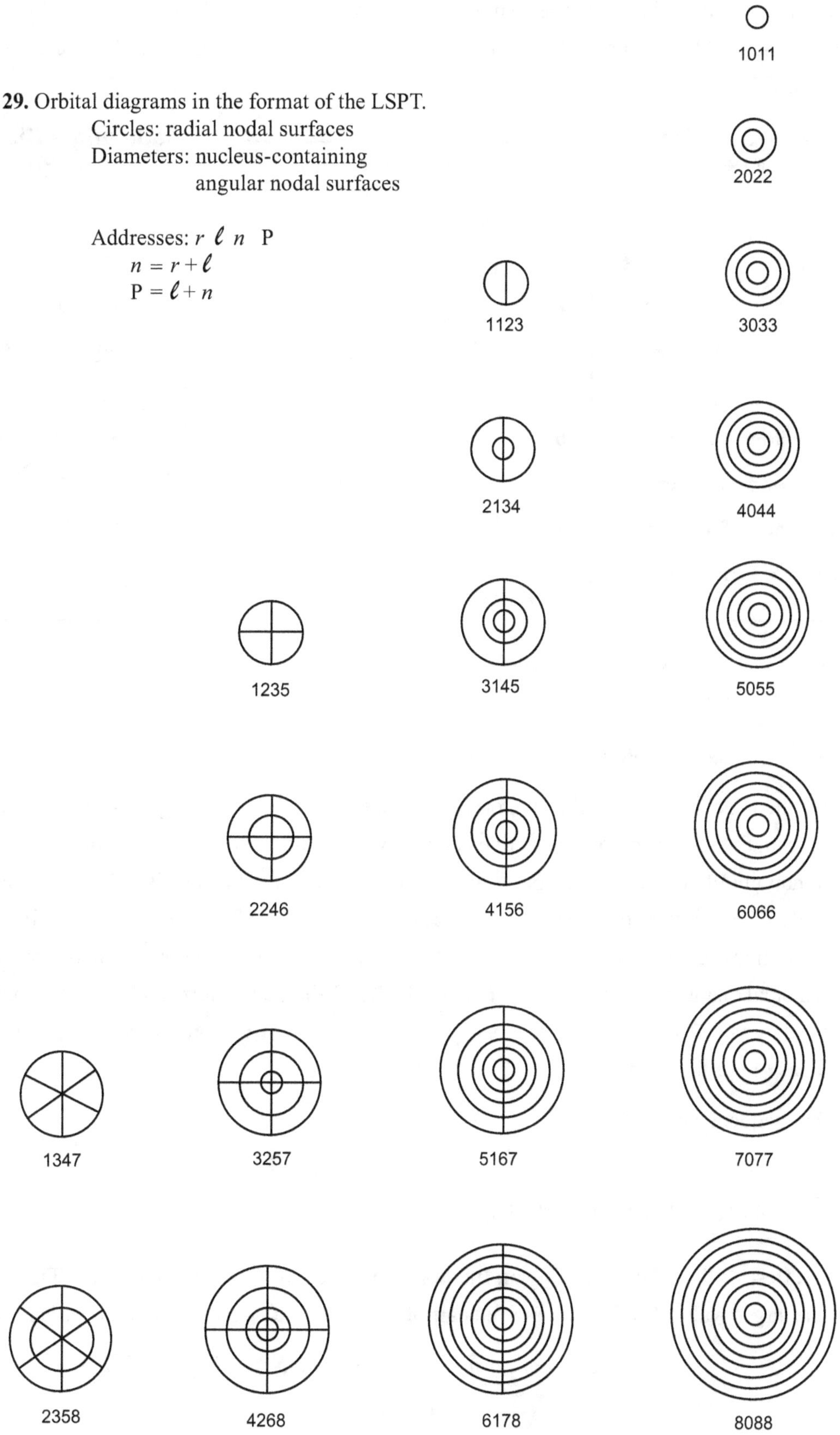

Replacement of the orbitals' circle-diameter representations by their "nℓ" labels yields what might be called the "LSPT (or Janet) Diagram". It's equivalent, mathematically, to Madelung's well-known diagram, Figure 30.

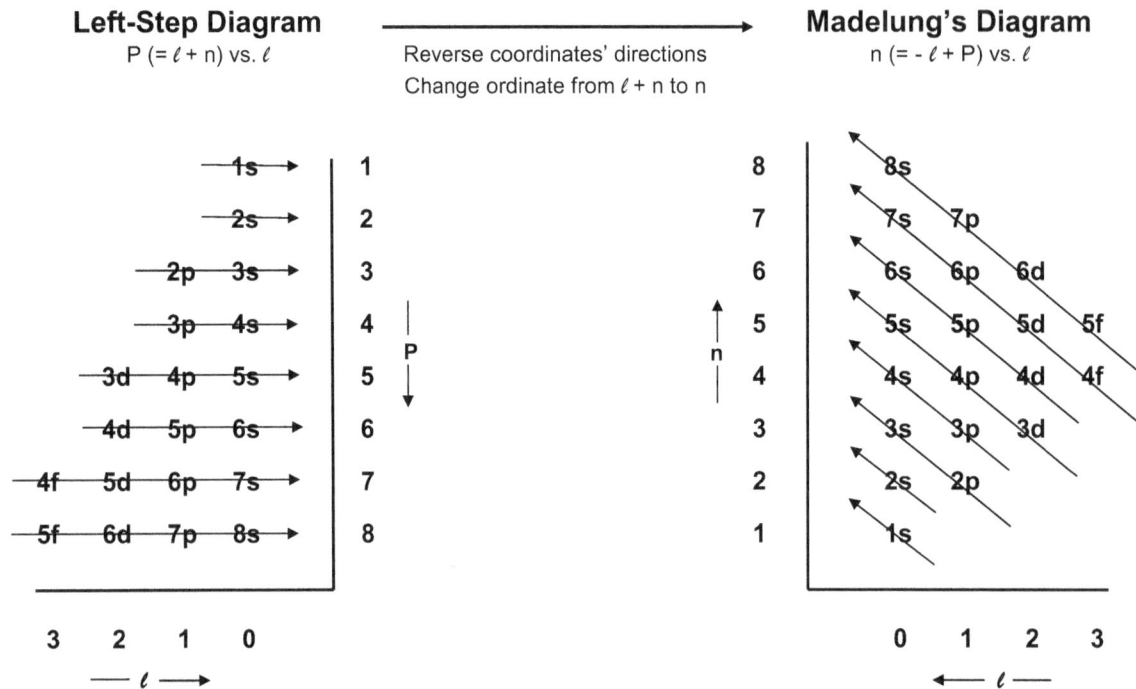

Figure 30. Relation of the LSPT to Madelung's diagram.

By the Block- and Row-Orbital Correspondence Principle, both diagrams are plots of the same thing: from the standpoint of descriptive chemistry, order of row occupancy in periodic tables; from the standpoint of atomic physics, order of orbital occupancy in atoms in Bohr's Aufbau Process. In the LSPT Diagram lines of constant n ($= P - \ell$) have slopes of +1: $P = \ell + n$(a constant). In Madelung's Diagram, lines of constant P ($= n + \ell$) have slopes of -1: $n = -\ell + P$(a constant). To obtain values of the principal quantum number n ($= P - \ell$) from the LSPT diagram, subtract abscissas ℓ from ordinates P. To obtain values of the orbital occupancy parameter P ($= n + \ell$) from the Madelung, diagram, add abscissas ℓ to ordinates n.

The sequence r, ℓ, n, P -

$$r \quad \ell \quad r + \ell \quad r + 2\ell$$

is the beginning of a Fibonacci series, Appendix XIII.

72. Description of the LSPT by an Inequality. Inspection of the left-hand-side of Figure 30 reveals that the largest value that ℓ has for a given pair of periods is the pair's ordinal number D minus 1.

An Expression of Form of the Dyadic Left-Step Periodic Table

$$\ell \leq D - 1$$

Large ℓ implies large D; i.e., late appearance of that value of ℓ in periodic tables. The inequality generates the following sets of values for ℓ.

D	1	2	3	4
ℓ	0	1 0	2 1 0	3 2 1 0

Dyad 1 contains rows only from block 0. Dyad 2 contains rows from blocks 0 and 1, dyad 3 rows from blocks 0, 1, and 2, and dyad 4 rows from blocks 0, 1, 2, and 3. Described, for ℓ-values increasing to the left, is a left-step table.

The corresponding inequality for the Madelung Diagram is the well-known expression $\ell \leq n - 1$. The inequality generates from n the following values for ℓ:

n	1	2	3	4
ℓ	0	1 0	2 1 0	3 2 1 0

Substitution for n from the relation $n = r + \ell$ into $\ell \leq n - 1$ yields $r \geq 1$.

The inequalities $\ell \leq n - 1$ and $r \geq 1$ have simple physical interpretations. Because all orbitals have a radial nodal surface at infinity, $r \geq 1$. Because the total number of nodal surfaces is $n = r + \ell$, $\ell = n - r$. For a given value of n, ℓ has a maximum value when r has its minimum value of 1. Hence $\ell \leq n - 1$: the number of nucleus-containing nodal surfaces, ℓ, is at most 1 less than the total number of nodal surfaces, n, owing to existence of a radial nodal surface at infinity.

Because D is, physically, the number of types of subshells occupied in atoms of dyad D, and because subshell occupancy occurs with increasing Z in order of increasing ℓ-values, ℓ's maximum value, if ℓ-values started at 1, would be D, but since ℓ-values start at 0, $\ell \leq D - 1$.

73. Shell-Filling. Figure 31 shows that shell n fills over a span of n physical periods.

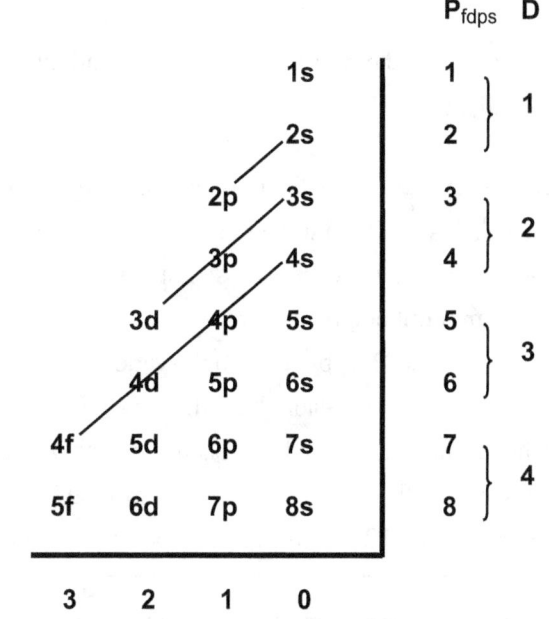

Figure 31. Shell-filling in Bohr's Aufbau Process.

Filling starts at the beginning of the *s*-block in period $P = n$ (= 4, e.g.) and ends at the end of the first occupied subshell of period $P = n + (n - 1) = 2n - 1$ (= 2x4 - 1 = 7).

Sometimes it's said that helium belongs above neon in periodic tables because, like neon, it has a full outer shell. Figure 31 shows that -

Neon is the only p^6 system that has a full outer shell.

Having a full outer shell is not a Group property for the Noble Gas Group. The fact that helium and neon atoms have only full shells is not grounds for placing helium in the Noble Gas Group.

Figure 32 illustrates the italicized statement immediately above. Displayed in the format of the LSPT are the principal quantum numbers, *n*, of the orbitals of the elements' differentiating electrons.

1 2 3 4 5 6 7 8 9 10 11 12 13 14 1 2 3 4 5 6 7 8 9 10 1 2 3 4 5 6 1 2

Figure 32. Principal quantum numbers of outer orbitals in the format of the LSPT.

```
                                                                          1  1
                                                                          2  2
                                                        2  2  2  2  2  2  3  3
                                          V             3  3  3  3  3  3  4  4
                   3  3 [3] 3  3  3  3  3  3  3  4  4  4  4  4  4  5  5
                   4  4  4  4  4  4  4  4  4  4  5  5  5  5  5  5  6  6
4 4 4 4 4 4 4 4 4 4 4 4 4 4 4  5 5 5 5 5 5 5 5 5 5 5  6 6 6 6 6 6  7 7
5 5 5 5 5 5 5 5 5 5 5 5 5 5 5  6 6 6 6 6 6 6 6 6 6 6  7 7 7 7 7 7  8 8
              f                          d                  p      s
```

$$V \quad 1s^2 2s^2 2p^6 3s^2 3p^6 4s^2 3d^3$$

* In the d- and f-blocks outermost *s-electrons* are the first electrons ionized. (+2 is a common oxidation number for Mn, Co, Ni, Cu, and Zn.)

The 1's in Figure 32 end at the end of the first row of the first dyad, at He. The 2's end at the end of the first row of the second dyad, at Ne.

74. The 2*n*² Rule. Figure 31 explains why the numbers of groups in Periodicity's periods are the same numbers $(2x1^2, 2x2^2, 2x3^2, 2x4^2)$ as the numbers of spin orbitals for different principal quantum numbers *n*, even though, for dyad D > 1, the orbitals associated with a single period in a periodic table have different principal quantum numbers, *n*.

Thus, the number of groups in dyad D is, by direct observation, $2D^2$. The number of spin orbitals of shell *n* is, by atomic theory, $2n^2$. The two expressions have the same algebraic form, although the subshells associated with dyad D and the subshells associated with shell *n* are, for D = *n*, different [with the exception (see the 4f orbital immediately below) that the first subshell of dyad D is the same as the last subshell of shell *n*]. Similarity of the algebraic forms $2D^2$ and $2n^2$ arises because, as one sees in Figure 31,

Atomic shell n and periods of dyad D = n run through the same ℓ-values.

Subshells of the First Period of Dyad 4				*Subshells of Shell 4*			
4f	5d	6p	7s	4s	4p	4d	4f

Because of the coincidence regarding numbers and types of subshells in Periodicity's dyads and atoms' shells, Pauli, in capturing his Exclusion Principle from Periodicity's "mystical numbers" 2, 8, 18, got the right answer for the wrong reason (Appendix II).

75. A Step Toward Correction of the Conventional Error in Conventional Periodic Tables.

The first step in analyzing the Helium Error in periodic tables is recognizing it for what it is: a category mistake. Helium's kinship with neon is not a *primary* kinship. It would seem, of course, that helium and neon must be related to each other in some fashion or other. They have the two highest ionization energies of any elements, the two lowest atomic refractivities, the two lowest proton affinities, the lowest and nearly the lowest electron affinities, and the first and third lowest boiling points. As elements go, they might seem to be as much alike as any pair of elements in the Periodic System. Also, their atoms are the only atoms in the entire System that have, as noted, full outer shells. Spectroscopically, however, helium and neon are not at all alike. Their atoms, concludes atomic physics, have entirely different electronic structures. And according to Mendeleev, in the Classification of the Elements according to the Periodic Law, data on the elements as atoms trumps data on the elements as simple substances.

The same remarks hold, in kind, for hydrogen and fluorine. As simple substances they are much alike: low boiling, covalently bonded diatomic gases that substitute for each other, atom for atom, in organic and inorganic compounds (Appendix VI). Yet, again, spectroscopically, in the visible and uv regions of their atomic spectra, the two elements are not alike. Accordingly their atom's electronic structures are not alike. And accordingly the two elements are not in the same Group in periodic tables.

> *Consistency in Periodic Tables calls for treating hydrogen and helium in the same manner.*

Either the Periodic Classification of the Elements is a classification of *atoms* or it is a classification of *simple substances*. In fact, in most instances, it's both. The two classifications generally coincide, *except* in the case of Nature's light elements, particularly hydrogen and helium. For hydrogen and helium, the two classifications yield entirely different results. A classification of atoms according to the Periodic Law places hydrogen above lithium and helium above beryllium in Periodic Tables, whereas a classification of simple substances according to the Periodic Law places hydrogen above fluorine and helium above neon. Accordingly, placement of hydrogen above lithium and *in the same table helium above neon is, for one or the other of the two placements, a category mistake.*

Before suggesting how to have one's cake (a kinship between helium and neon) and eat it too (no category mistake), this report pauses briefly to comment, again, on a leading challenge for theoreticians: Provide a mathematically rigorous and physically comprehensible explanation of the Row-Orbital Order of Occupancy Rule: smallest $n+\ell$ first and, for a given $n+\ell$, largest ℓ (or, as it's often said, smallest n) first. (The proviso "smallest n first" is not useful in description of the top table in Figure 27.)

76. Logical Status of Madelung's Rule. Madelung's Orbital-Occupancy Rule was based, in the first instance, on spectroscopy's empirical classification of atoms' term values and chemistry's empirical Periodic Classification of the Elements. It is supported by a Row-Orbital Correspondence Principle. It can be accounted for, qualitatively, in terms of electron screening and penetration. It has been confirmed by quantum mechanical calculations. But it has not yet been deduced from first principles in a physically meaningful manner (37).

At the present time, no theorem that a *mathematical* proof, based solely on physical principles, of the Madelung Orbital Occupancy Rule exists. [Chemical proof of the Rule's chemical significance (the left column of Table 1) is summarized in Figure 1. Well-known quantum mechanical results yield the physical side of the Correspondence (cited on the right in Table 1).] In fact, an analysis of values of k in the Madelung form $(n + k\ell)$ required by the LSPT (**69**) and consideration of the dependence of the order of atoms' excited states on their atomic numbers (Figure 28) suggest that no such proof exists (**69**). In that event, the argument that physics has not explained the periodic table because it has not produced an analytical proof of the Madelung $(n+\ell)/\ell$ rule naively faults physics for not doing the impossible. Neither, however, does a rigorous proof of a non-existence theorem exist. Inasmuch as *verbal* expression of the Madelung Rule in terms of Madelung's parameter expressed in the form $r + 2\ell$ suggests a *verbal* rationalization of the Rule, so, also, one might wonder: Might an *analytical* expression of the Madelung Rule suggest an *analytical* account of the Rule?

77. A Row-Orbital Order-of-Occupancy Equation. The diagrams of Janet and Madelung (Figure 30) generate a correspondence between the ordinal numbers R of blocks' rows in Periodic Tables and atomic orbital descriptors $n\ell$, shown in the first two lines of Figure 33.

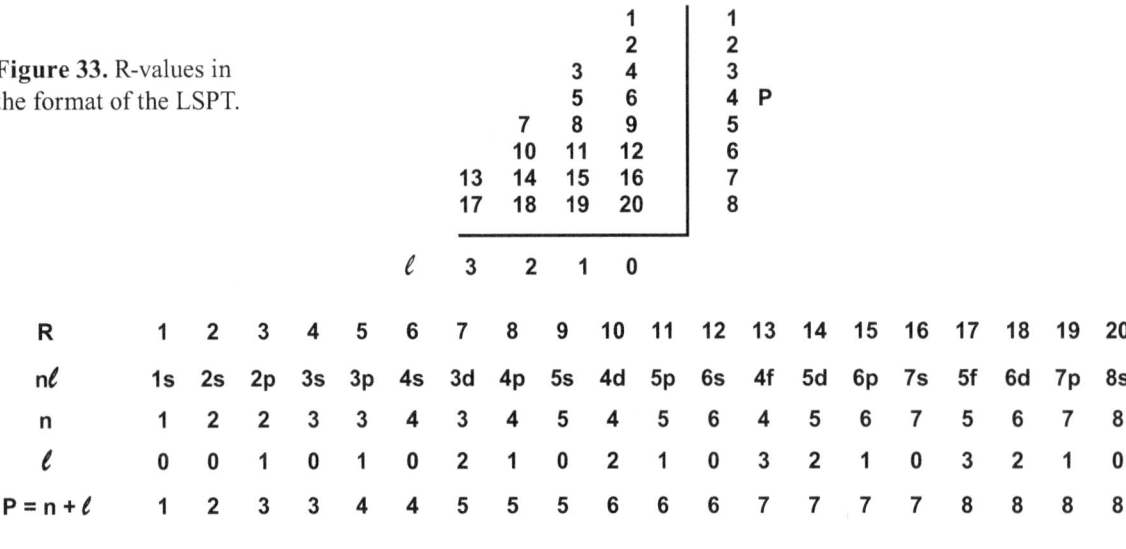

Figure 33. R-values in the format of the LSPT.

R	1	2	3	4	5	6	7	8	9	10	11	12	13	14	15	16	17	18	19	20
$n\ell$	1s	2s	2p	3s	3p	4s	3d	4p	5s	4d	5p	6s	4f	5d	6p	7s	5f	6d	7p	8s
n	1	2	2	3	3	4	3	4	5	4	5	6	4	5	6	7	5	6	7	8
ℓ	0	0	1	0	1	0	2	1	0	2	1	0	3	2	1	0	3	2	1	0
$P = n + \ell$	1	2	3	3	4	4	5	5	5	6	6	6	7	7	7	7	8	8	8	8

$$R = (P^2 + 2P + \delta)/4 - \ell \quad \delta = 0 \text{ or } 1 \text{ for P even or odd}$$

The bottom two lines of Figure 33 (labeled "ℓ" and "P") illustrate Madelung's Rule: "Smallest P $(= n + \ell)$ first [on moving along the P line left-to-right] and, for a given P, largest ℓ first."

The *n* row beneath Figure 33 illustrates the rule that the Index of Discontinuity, *n*, adopts a new value on each entrance to the *s*-block.

The equation at the bottom (derived in Appendix XIV) reproduces the Madelung Rule and the associated features of the Left-Step Periodic Table, Table 6.

Table 6. Implications of the R-equation.

o   For given P, the row occupied first (smallest R) has the largest ℓ.

o   Large ℓ implies large P (since R ≥ 1).

o   As ℓ increases, $P_{(fdps)}$ constant, R decreases by the same amount:
$$(\Delta R/\Delta \ell)_P = -1.$$

o   As P increases, ℓ constant, R increases by the amount (taking into account the role of δ) given by the expression:
$$(\Delta R/\Delta P)_\ell = (P + 2 - \delta)/2$$

Values for the ratio reflect Periodicity's dyadic character.
For P = 1 – 7, the ratio's numerical values are 1 2 2 3 3 4 4.

High hopes for the R-equation vanish quickly. It does not go beyond its input. The quadratic dependence of R on P merely reflects Periodicity's dyadic character. Illustrated is a fundamental feature of logic: Output equals input. The R-equation is nothing more, nor less, than an algebraic restatement of the geometrical shape of the LSPT. Its derivation requires no fresh physical input (Appendix XIV).

P-values in the R-equation are ordinal numbers of the LSPT's periods. The substitution $P = n + \ell$ yields an expression for R in terms of *n* and ℓ applicable to *all* periodic tables. The further substitution $n = r + \ell$ yields an expression for R in terms of *r* and ℓ. The expressions become increasingly complex. Illustrated is the fact that in the company of the radial, angular, and principal quantum numbers, *r*, ℓ, and *n*, the Madelung Parameter $n + \ell$ $(= P_{fdps} = P_{LSPT})$ represents, so to speak, "the *natural* quantum number" of *many-electron atoms*. Their orbitals' energies do not depend on P in the simple fashion that hydrogen atoms' orbital energies depend on *n*, but they are ordered by values of P $(= n + \ell)$, and ℓ (or, simply, by, e.g., $n + 0.9\ell$).

The R-equation, stated in terms of *r* and ℓ, expresses analytically the conclusion (reached topologically in Section **100**) that all periodic tables are, in a deep sense, equivalent to each other. In the order of occupancy of their blocks' rows with increasing atomic number they are identical, while, nonetheless, differing in their blocks' rows' physical arrangements in sometimes chemically significant — and complementary — ways.

78. Plan B: Recourse to the Anthropic Principle. For proof of Madelung's Rule, one is led to fall back on the thought that if Periodicity obeyed a different rule, chemistry would be different and mankind might not be here to talk about it. Presence of human beings in the universe may demand the $(n+\ell)/\ell$ Row-Orbital Occupancy Rule. For suppose, for example, that electron screening-and-penetration effects in many-electron atoms were smaller than they are. The Madelung Rule might be: smallest n first and, for given n, smallest ℓ first. The periodic table of the chemical elements would be the shallow, right-step table of Figure 27, repeated in Figure 34 below.

```
 1   2
 3   4   5   6   7   8   9  10
11  12  13  14  15  16  17  18   19  20  21  22  23  24  25  26  27  28
⎵⎵⎵⎵⎵⎵   ⎵⎵⎵⎵⎵⎵⎵⎵⎵⎵⎵⎵⎵⎵⎵⎵⎵⎵   ⎵⎵⎵⎵⎵⎵⎵⎵⎵⎵⎵⎵⎵⎵⎵⎵⎵⎵⎵⎵⎵⎵⎵⎵⎵⎵⎵⎵⎵⎵⎵
     s              p                                 d
```

Elements of the *d*-block would start with element 19, not 21. Atoms of the first d^6 system, our iron, would have a nuclear charge of 24, not 26. That change in nuclear charge might destroy iron's finely-tuned capacity to bind and release oxygen in hemoglobin in a manner supportive of life as we know it.

Of course, the argument's too easy? One might prove, thereby, almost anything? Including a theorem for which a proof may be nonexistent (**69, 76**)?

79. Secondary Chemical Kinships and the International Column-Labeling Controversy. Periodic tables' primary purpose is exhibition of chemical kinships, primarily *primary kinships,* i.e., kinships among elements that are in the same Group, usually indicated by verticality. Periodic tables' second leading purpose is exhibition of *secondary kinships*, i.e., kinships among elements that are in different Groups (in different blocks) that are the same distances from their blocks' left ends *and that exhibit the same maximum states of oxidation.* Leading examples involve elements of the second rows of the *p*- and *d*- blocks with, respectively, elements of first rows of the *d*- and *f*-blocks.

Leading Examples of Secondary Chemical Kinships

p d: Al Sc   Si Ti   P V   S Cr   Cl Mn   *d f*:   Y La   Zr Ce

Because the first Group of the *s*-block is the only Group that has a maximum oxidation state of +1, its elements participate in no secondary kinships.

The alkaline earth metals (Be Mg Ca Sr Ba Ra) and the volatile metals (Zn Cd Hg) have the same maximum oxidation numbers (+2). However, since they are not the same distances from their blocks left-ends, their elements are not, by the previously cited definition, related by secondary kinships.

Traditionally, secondary chemical kinships have been displayed graphically by tie-lines in step-pyramid tables and by adjacency in "short-form" tables, Figures 35 (following page).

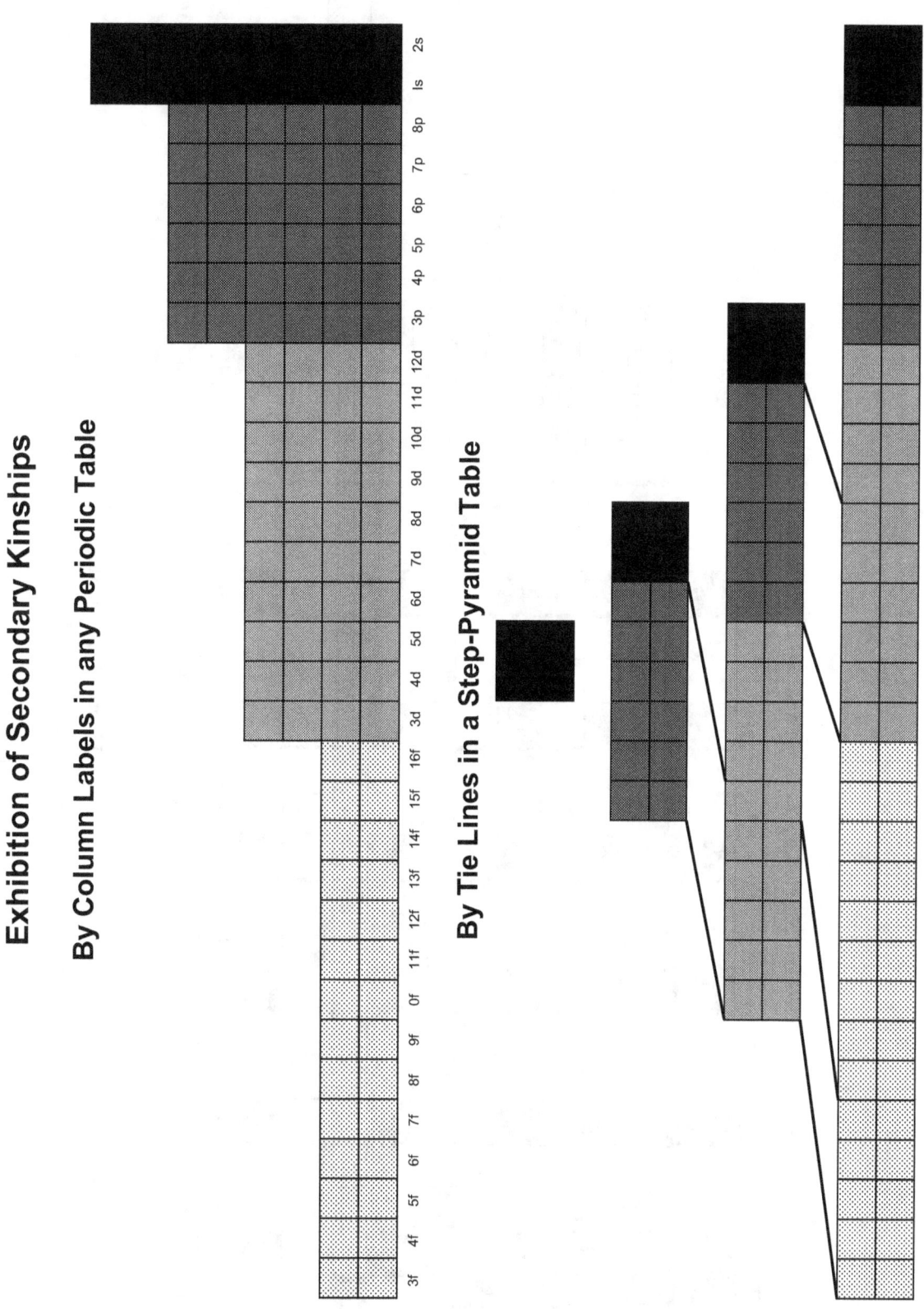

Figure 35. Different ways of exhibiting secondary chemical kinships.

By Proximity in a "Short Form" Table

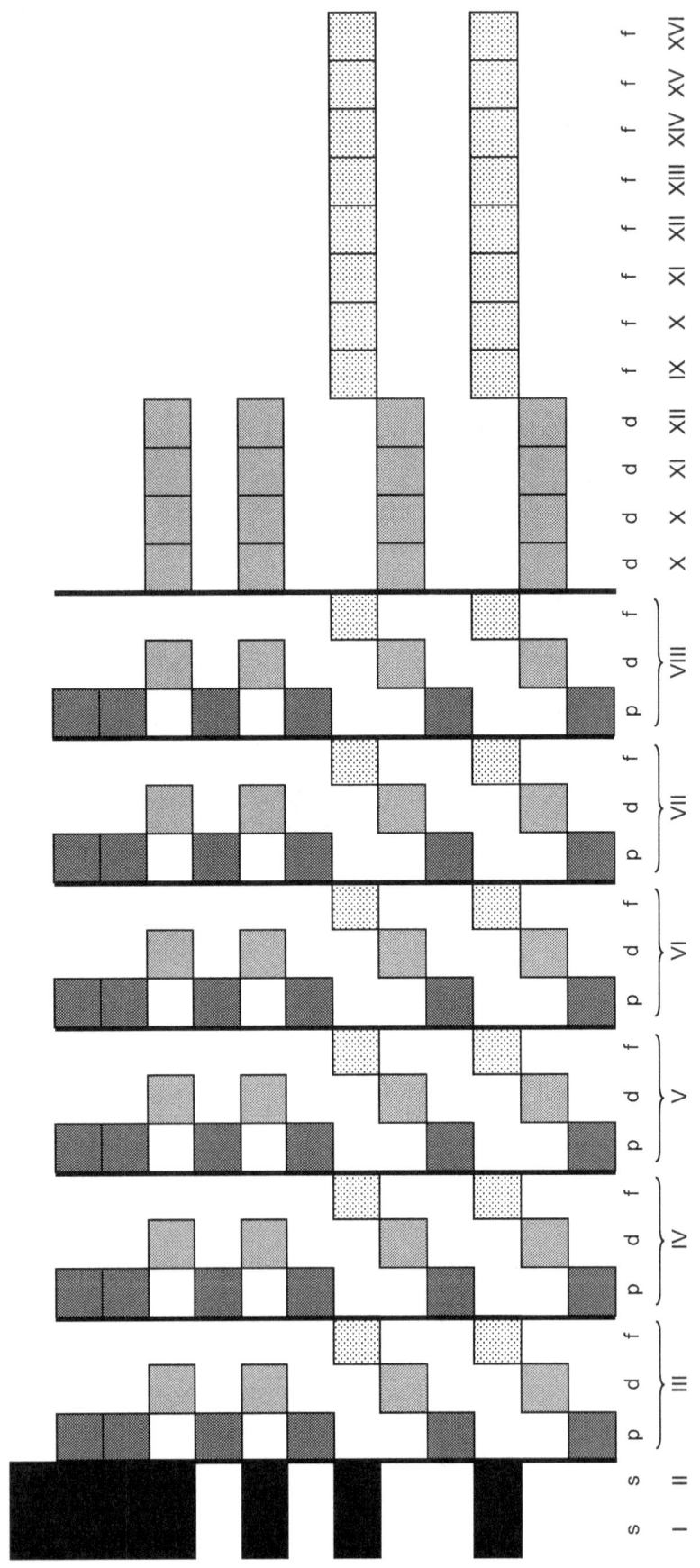

Figure 35 continued.

Each graphic expression of secondary chemical kinships requires relaxation of one or more of the Construction Conventions cited for Figure 1. The step-pyramid table (second table of Figure 35) sacrifices verticality of primary kinships, along with dyads' adjacency. [The usual step-pyramid table has the *s*-elements on the left and, consequently, has at the top, instead of a dyad, a monad, useful for indicating by tie-lines tertiary kinships of H and He with, respectively, F and Ne (**80**]. The modern version of Mendeleev's favorite table (last table of Figure 35) has interrupted columns, interrupted rows, and interrupted periods.

The three tables of Figure 35 are the same width. Each table has 32 columns. In descriptions of periodic tables, the traditional phrases "long form", "medium-long form", and "short form" betray fundamental misunderstandings of the Periodic System.

Ready construction from step-pyramid tables of Jensen family trees, such as the one for Family IV shown in Figure 36, below, is one of the attractive features of step-pyramid tables.

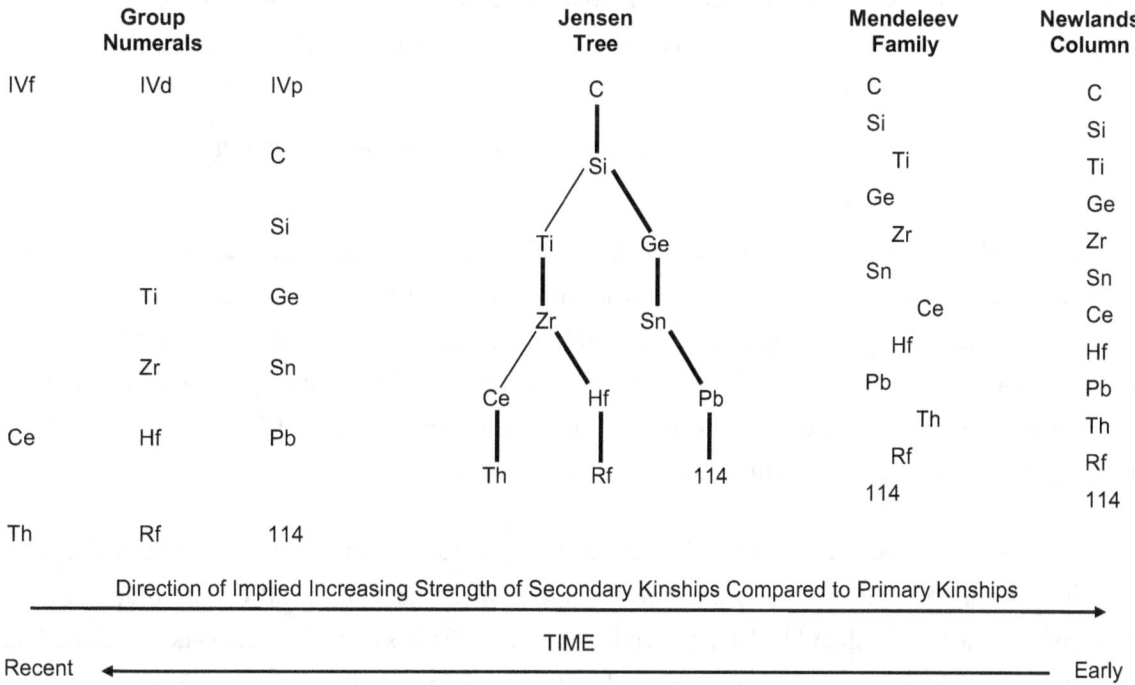

Figure 36. Four displays of Family IV, after Newlands, Mendeleev, and Jensen.

Verticality and heavy tie lines in Figure 36 indicate primary kinships, except in Newlands' column at the right, which displays in a single, vertical column all elements in the Periodic System that have a maximum oxidation number of +4 (following elements that have a maximum oxidation number of +3). Off scale to the left, and not shown, is IUPAC's 1-18 scheme of column labels, which do not acknowledge existence of secondary chemical kinships. One of the major achievements in the history of the Periodic System has been separation of the elements of Families, such as those of Family IV (consisting of the C, Ti, and Ce Groups), into three Groups of elements, each of which (it's turned out) is characterized by a distinctive type of differentiating electrons (*p*, *d*, or *f*).

Mendeev's display of a "Mendeleev Family" is graphically less elegant than the corresponding "Jensen Tree" but conceptually complementary. Mendeleev's columns within a family contain, from the left, *p*-, *d*-, and *f*-type elements. Light tie lines of a "Jensen Tree" indicate secondary kinships, between *p*- and *d*-elements and between *d*- and *f*-elements. Figures 35 and 36 illustrate how the phenomenon of secondary kinships has given rise to different periodic tables — some 700, and counting.

Secondary kinshipness holds only for elements in their highest states of oxidation. "[T]he resemblances of the properties of the higher form [of oxidation] does not extend to the lower forms," writes Mendeleev (38), "and even entirely disappears in the elements, for there is not the smallest resemblance between [e.g.] sulphur and chromium . . . in the free state."

Existence of secondary kinships has led to: creations of "short-form" and step-pyramid periodic tables; Mendeleev's discussions of of "even" and "odd" series; alphanumeric column labels; an international controversy regarding use in such labels of the suffixes "A" and "B"; formation of an American Chemical Society Presidential Ad Hoc Advisory Committee on Column Labels (Appendix XXII); attempts by the Committee's chairman to discover logical (rather than political) grounds for resolving the column-labeling controversy; and, after two decades, this report, and the conclusion, here stated for the first time, that -

The international column-labeling controversy was a tempest
in a teapot.

created by (i) the unstated assumption on the part of all concerned parties [IUPAC (the International Union of Pure and Applied Chemistry), European chemists, American chemists, chemical educators, chemical abstracts' computer programmers, and national chemical society nomenclature committees] that everyone should use, 24/7, in all situations, the same column labels; and by (ii) IUPAC's implicit assumption that everyone did, and would continue to use, for all time, 24/7, in all situations, the currently Conventional, "18-Column", *s*(*f*-footnoted)*dp* Periodic Table.

80. Tertiary Chemical Kinships. Jensen has recognized, and named, a third type of chemical kinships: tertiary kinships. Tertiary kinships involve elements that are in different Groups that are the same distances from their blocks' *right* ends and that have the same *minimum* states of oxidation. Leading examples are hydrogen & fluorine and helium & neon, Figure 37 (following page).

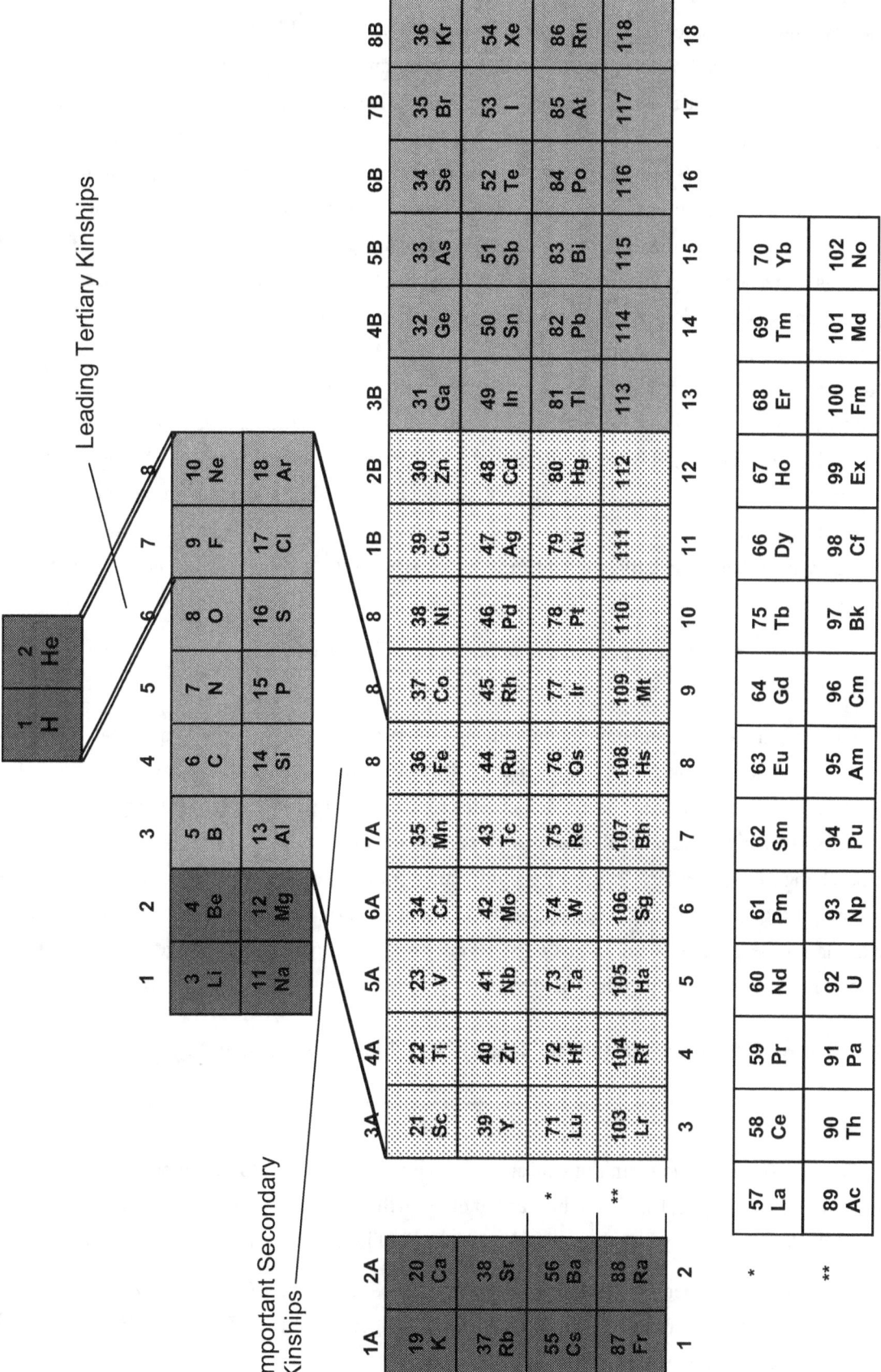

Figure 37. A hybrid step-pyramid/block-type periodic table.

73

The periodic table of Figure 37 was popular for many years with inorganic chemists, particularly in Europe. Pauling used it because (private communication) he did not like to see magnesium separated from aluminum. Association of adjacency with similarity and similarity with adjacency is one of the Periodic System's most widely known features. Major exceptions are summarized in the Mendeleev Rule of Light-Element Distinctivenes (30). In the original European AB column-labeling scheme, shown in Figure 37, suffixes A and B are not applied to elements in the table's second and third periods.

The AB column-labeling scheme of Figure 37 was designed for periods that exhibit *two* runs through maximum states of oxidation, i.e. the sdp periods 4 & 5 of Conventional Periodic Tables. It does not, as mentioned, apply to the sp periods 2 & 3, which exhibit only *one* run through maximum oxidation numbers. Nor does the European AB scheme apply to *sfdp* periods 6 & 7, which exhibit *three* runs through maximum states of oxidation.

Noteworthy are tertiary kinships involving two elements at the end of the *d*-block and, again, the last two groups of the *p*-block: namely, between (i) gold (a "pseudohalogen", as, e.g., in cesium auride) and the halogens, and between (ii) mercury (a "pseudo-noble gas") and the Noble Gases. No tertiary-type kinships involving *f*-block elements have, to the author's knowledge, been cited in the literature. Like, however, the volatile metals at the end of the *d*-block, ytterbium at the end of the *f*-block is by far the most volatile element in its block. Broadly speaking, Nature appears to exhibit the following -

Block/Block Trend in Strength of Tertiary Kinships

$$s/p \; > \; p/d \; > \; d/f$$

The strong (if atomically distant) helium-neon *s/p* tertiary kinship helps to domesticate the weak (but atomically close) helium-beryllium kinship. Movement of s^2 He from its conventional location above p^6 neon in the *p*-block to the *s*-block creates a new, if nature's weakest, primary kinship and, simultaneously, a new, and nature's strongest, tertiary kinship.

The change He/Ne to He/Be is, in its significance for the Periodic System, akin to Lavoisier's change of "phlogiston lost" to "oxygen gained". Lavoisier's correction of traditional thought did not deny a relation between a metal and its calx. It redefined it. Similarly, the change He/Ne to He/Be does not deny a relation between helium and neon. It redefines it.

81. Electronic Characterization of Chemical Kinships. Jensen defines kinships electronically (39), Table 7.

Table 7. Jensen's electronic definitions of chemical kinships, designed to accompany an sfdp, step-pyramid periodic table (alkali and alkaline earth metals on the left-hand side) with (consequently) H and He in a "block" by themselves at the top of the table, as in Figure 37, a modified step-pyramid table.

> <u>Primary or Isovalent Relationship</u>: *identical valence electron and valence vacancy counts.* By the latter part of that definition, H is *not* a congener of Li *nor* He of Be. Zn, on the other hand, is a congener of *both* Ca and Zn.

Secondary or Isodonor Relationship: *identical valence electron count, different valence vacancy count.* By that definition, H has a secondary relationship with Li, and likewise for He with Be.

Tertiary or Isoacceptor Relationship: *identical valence vacancy count, different valence electron counts.* By that definition, H has a tertiary relationship with F, and likewise for He with Ne.

By those definitions, hydrogen and helium participate in secondary and tertiary kinships, but not in primary kinships.

"Identical valence electron counts" means "same distance from blocks' left ends" for elements in the *p-*, *d-*, and *f*-blocks. At blocks' outsets, it means, also, "same maximum oxidation numbers". "Identical valence vacancy" means "same distance from blocks' right ends". Near those ends, it means "same minimum oxidation number". "Identical valence electron *and* identical valence vacancy counts" means "in the same group". "Identical valence electron count and different valence vacancy count" and "identical valence vacancy count, different valence electron counts" mean "in different groups in different blocks".

The two sets of kinship definitions, phenomenological (or chemical), based on the shapes of periodic tables (Sections **79** and **80**), and physical (or electronic), based on atomic structure (Table 7) agree that *the helium-neon relationship is not a primary relationship.* Nonetheless, they yield, in detail, different results in three instances: H, He, and Zn.

The question Which set of definitions is best, or right? misses the point. "Right" or "best" for which Periodic Table? A Step-Pyramid Table, s-elements on the left, H and He above all dyads, in a monad by themselves? Or a Left-Step Table?

The *sfdp*, Step-Pyramid Periodic Table allows one to draw a primary kinship tie-line, if one so wishes, between zinc and magnesium and to include, if desired, the Zinc Group in a "Main Block" of *nine* Groups that have elements in the *sfdp* periods 4–7 (compared to only eight Groups in periods 2 and 3). The "transition block" is left, accordingly, with nine Groups (instead of the usual ten). Gained is graphic expression of Jensen's definition of primary relationships, applied to Zn and Mg. Lost are regularities of Periodicity that depend on location of the volatile metals in the *d*-block (i.e., most of the regularities reported in this report). An additional reason for preferring *in this report* (on uses of the LSPT) phenomenological definitions of chemical kinships is cited in Section **89**.

DISCLAIMER: This report does not claim that the phenomenological kinship definitions that place hydrogen and helium at the top of the *s*-block are "right", only that "This is what happens when one proceeds in that manner": emergence of regularities absent when helium is located above neon.

82. Quaternary Kinships? Electronic characterization of different types of chemical kinships by the words "isodonor" (same number of valence-shell electrons) and "isoacceptor" (same number of valence-shell vacancies) raises the question: Are their kinships between atoms that are, to coin a word, "isodonoracceptor"? Are there, in other words, kinships between atoms whose valence electron counts are equal to their valence electron vacancies? Are there kinships between atoms whose valence shells are *half full?*

The first such atom is hydrogen, of the *s*-block. In the *p*-, *d*-, and *f*-blocks, where outer *s*-electrons are valence shell electrons, the total number spin-orbitals in an atom's valence shell is $2 + 2(2\ell + 1)$. Half that number is $2\ell + 2$. Since two of the half-valence-shell-filling electrons are *s*-electrons in the *p*-, *d*-, and *f*-blocks, the number of electrons in their valence shells' ℓ subshells is 2ℓ. The atoms in question are, in short, the atoms with the outer subshell configuration -

$$\boldsymbol{\ell}^{2\ell}$$

where the bold-face $\boldsymbol{\ell}$ stands for *p*, *d*, or *f* and the exponential ℓ stands for, correspondingly, 1, 2 or 3. The atoms in question are, briefly put, p^2, d^4, and f^6 systems. First examples in the *p*-, *d*-, and *f*-blocks are C, Cr, and Sm.

And indeed: hydrogen's *chemical* location in periodic tables, it has recently been suggested, in part because of its half-filled shell, is above carbon (16). The two elements, H and C, have, e.g., the strongest and second strongest single bonds in chemistry. And, in the *d*-block, chromium's congener tungsten has the highest heat of vaporization of any element. In the *f*-block, however, the element with the highest heat of vaporization is thorium, not samarium. The quaternary kinship concept appears to be of limited usefulness. Additionally, an across-the-table, phenomenological definition for it, applicable to all blocks, does not appear to exist.

83. Similarity and Adjacency, "Strength" and "Closeness". Jensen's kinship distinctions and terminology are useful in statements of additional distinctions, Table 8.

Table 8. Similarity and Adjacency in Block-Type Periodic Tables.

Elements may be –

- similar and adjacent, vertically. Examples are primary kinships between heavy elements.

- similar and adjacent, horizontally. Examples include the rare earths and the corresponding actinides and, to a lesser extent, the light and heavy platinum metals.

- not similar but adjacent, vertically. Leading examples are primary kinships that involve light elements.

- not similar but adjacent, horizontally. Leading examples are elements at the LSPT's *s*-block/*p*-block interface.

- similar but not adjacent. Leading examples are a strong special kinship between Mg & Zn and strong tertiary kinships between He & Ne and H & F.

- not similar and not adjacent. Examples are most pairs of elements in periodic tables.

Kinship *strength* is not necessarily related to kinship *closeness*. "Strength" in the present context refers to similarities of elements as *simple substances*. "Closeness" refers to similarities of elements as *atoms*. Primary H/Li and He/Be kinships are close, as regards types of atoms, but weak, as regards properties of the corresponding simple substances. On the other hand, tertiary H–F and He–Ne kinships are distant, as regards types of atoms, but strong, as regards properties of the corresponding simple substances.

Historically, many mistakes in periodic tables have been secondary and tertiary chemical kinships mistaken for primary kinships — even to this day (in the case of helium). Other similarities that have led, temporarily, to incorrect kinship assignments include: Tl (owing to its common oxidation state Tl^+) and Cu, Ag, and Au (as, similarly, monovalent cations) with the alkali metals; Pb (as Pb^{+2}) with the Alkaline Earth Metals; Cr (as Cr^{+3}) with Al; and the rare earth metals (as M^{+3}) with Sc and Y.

84. Non-Traditional Triads. Secondary and tertiary kinships yield secondary and tertiary triads. Two members of such triads belong to the same Group. The third element does not. In secondary triads, the odd element out is the *third* member of the triad. Ramsay has cited four such triads (40): F Cl Mn; O S Cr; N P V; and C Si Ti. They might be designated ppd triads. Another secondary triad is the ddf triad Sc Y La. In tertiary triads, the odd element out is the *first* member of the triad. Leading examples are the spp triads H F Cl and He Ne Ar. In all the examples of non-traditional triads cited, the triads' first members are *first-row elements*.

The only first row elements that form triads with the first two elements of the next block of larger ℓ are H and He. Ne, e.g., does not form a triad with Zn and Cd. Nor does Zn form a triad with Yb and No. Helium, as noted, forms a triad with Ne and Ar, and H with F and Cl, because the s-block is unique. It is the only block in the Periodic System where the first exit from the block (from He) is back to the block itself (to Li).

Mg forms with Zn and Cd a special sdd triad. Na forms a similar triad with Cu and Ag.

The more periodic tables exhibit *by verticality* strong, non-primary triads, such as He above Ne and Ar, Al above Sc and Y, Mg above Zn and Cd, Na above Cu and Ag, and Sc and Y above La, the more the tables' shapes depart from the perfectly regularity of the Left-Step Periodic Table. Usually all of those misassignments have been eliminated from periodic tables, except the first one.

85. Estimates of Helium's Boiling Point. Elements of the tertiary kinships H-F and He-Ne are, as mentioned, analogous to each other as simple substances. The analogy H is to F as He is to Ne yields the following estimate for helium's boiling point:

$$H/F = He/Ne$$

➔

$$He \approx Ne(H/F) = 27.1\ K(20.2/85.0) = 6.5\ K \quad (Obs.: 4.3\ K)$$

The estimate looks more impressive expressed on the Celsius scale. Substitution of normal boiling points on that scale, plus "x", into the expression H/F = He/Ne yields for "x" the value 277.6. An analogy based on Periodicity's leading tertiary kinships yields an estimate of the absolute temperature of water's triple point accurate to within two percent.

The ratio of the ratio H/F to the ratio He/Ne may be written as the ratio of H/He to the ratio of F/Ne. The approximate equality of the ratio of the ratios of the boiling points of the last two elements of the first rows of the blocks of outer-outer and outer orbitals, (H/He)/(F/Ne), to the same ratio of ratios for the blocks of inner and inner-inner orbitals, (Cu/Zn)/(Tm/Yb), yields the following estimate for helium's boiling point:

$$
\begin{aligned}
\text{He} \quad &\approx \quad \text{H x (Ne/F) x [(Tm/Yb)/(Cu/Zn)]} \\
&= \quad 20.3 \text{ K x (27.1/86.0) x [(2220/1466)/(2840/1180]} \\
&= \quad 4.01 \text{ K}
\end{aligned}
$$

86. Mendeleev's "Absolute Distinction" Revisited. The previous section presents what might appear to be, on the face of it, a logical inconsistency. Emphasized in this report has been Mendeleev's statement that the Periodic System is about atoms, not simple substances. On the other hand, the previous section uses a tabular expression of the Periodic System to create what might be called, after Mendeleev, a "simplesubstanceanalogy", for estimating a normal boiling point, not normally considered to be an atomic property. Of course, in principle, boiling points are *explainable* in terms of atomic properties, and, therefore, might be considered to be, so to speak, secondary atomic properties, perhaps especially in the case of the monatomic elements He and Ne and, slightly less so, for the diatomic elements H and F.

The word *explainable* (as used immediately above) expresses in a nutshell the fact that the relation to each other of properties of atoms and properties of simple substances is asymmetrical, obviously. "The simple substance diamond is composed of carbon atoms," chemists say. They do not say "Carbon atoms are composed of diamond." Properties of atoms, chemists say, are more *fundamental* than properties of simple substances. Passage from a carbon atom's electronic structure to diamond's hardness is a credible deduction from plausible principles, whereas passage from diamond's hardness to a carbon atom's electronic structure would be deemed to be an incredible induction from implausible leaps of imagination.

The distinction between the properties of atoms and properties of simple substances resides in the theoretical structure of chemistry. Mendeleev's statement that the Periodic System is about atoms, not simple substances, is a short-hand statement regarding the nature of chemical thought — which, chemists like to think, constantly improves. (It was a chemist, Lavoisier, the author recalls reading, who, in a description of chemistry, first used the word "progress" to mean "steady improvement".) Not surprising, therefore, is an occasional need to update schemes — such as periodic tables — that are based on chemical thought.

Two things *are* surprising about Mendeleev's classification of the chemical elements according to the Periodic Law. One is how much of the classification has not needed revision. The other is how long the scheme's second element has been misplaced, above neon, despite Mendeleev's Rule of Light-Element Distinctiveness; despite his statement that the Periodic System is about atoms, not simple substances; despite classification by atomic physics of helium atoms as s2 systems (not p6 systems); despite appearance of the Left-Step Periodic Table nearly eight decades ago; and despite numerous implications of the LSPT that require placement of helium above beryllium. In retrospect, it almost seems that, based on his methods of reasoning, Mendeleev could have placed helium in the Periodic System correctly at the outset.

87. Atomic Weights and Helium's Location in Periodic Tables. The most fundamental property of an element in Mendeleev's view is its atomic weight (41). The simplest, Mendeleevian "atomanalogy" involving that atomic property for helium is this: the atomic weight of helium's immediate successor in the Periodic System, Li, is to the atomic weight of helium as helium's atomic weight is to that of its immediate precursor, say X:

$$Li : He :: He : X$$

Replacement of the ":" of "is to" by "−" and the "::" of "as is" by "=" yields -

$$Li - He = He - X$$
$$\rightarrow$$
$$X = He - (Li - He) = 4 - (7 - 4) = 1$$

Helium's immediate precursor in the Periodic System is hydrogen. There are not, as Mendeleev sometimes supposed (owing to location of helium above neon), any elements between helium and hydrogen in the Periodic System.

If the phrase "no elements between X and Y in the Periodic System" means "no gaps between X and Y in the periodic table", as is the case, e.g., in Mendeleev's table of "The Periods of the Chemical Elements" (except for its first period, H - - - - - - He), then helium's adjacency to hydrogen in the Periodic System implies that helium's location in periodic tables is above beryllium.

If beryllium has a lighter congener, Y, then the "atomanalogy"

$$(Li - H)/(Na - Li) = (Be - Y)/(Mg - Be)$$

yields for Y's atomic weight the expression

$$Y = Be - [(Li - H)/(Na - Li)](Mg - Be)$$
$$= 3.3$$

Locations of H, Be, Li, Na, and Mg in periodic tables and their atomic weights imply that if beryllium has a congener lighter than itself, its atomic weight is probably about 3.3.

One of Mendeleev's leading achievement in chemistry was his use of atomic weights to state the Periodic Law and his use of the Law to (i) correct atomic weights and to (ii) predict, by "atomanalogies", the existence and properties of several undiscovered elements. His chief failures, regarding the Periodic System, were, in retrospect, (i) overlooking use of the Periodic Law and atomic weights to predict the existence of an *entire Group* of elements, the Noble Gases; and (ii) overlooking use of "atomanalogies" and his rule of light-element distinctiveness in placement of helium in his periodic tables.

Mendeleev might have further rationalized location of helium above beryllium by viewing it, along with fluorine (and, later, neon), as an example of inertness toward oxidation of the last elements in his Periodic System's first two "series": H He and Li Be B C N O F (Ne). By several lines of thought, therefore, Mendeleev might have rationalized placement of helium above beryllium in periodic tables. Of course, the "Mendeleev" of the previous remarks stands for someone with 20/20 hindsight. Hindsight leads, also, to a practical objection to placement of helium above neon.

88. A Philosophical Impracticality: He/Ne. Implicit in the character of chemical thought is a practical reason for not locating helium above neon in periodic tables.

> *The most fundamental and, thus, in the long run, the most useful classification of the elements is in terms of those properties of the elements that are the most explanatory.*

Helium's inertness and low boiling point are not as explanatory as its atomic structure.

Of course, placement of helium above neon in a tabulation in which, generally speaking, adjacency implies similarity, does, indeed, yield a simple account of helium's most familiar properties (inertness and volatility) vis-à-vis the corresponding properties of neon. One notes, however, that -

> *Helium-above-neon is a pre-atomic-structure classification.*

It dates back to 1900. And, so? That's an argument against He/Ne? Isn't most of the Periodic Classification of the Elements as worked out by Mendeleev from the facts of chemistry before 1900 considered valid today? Yes, of course, except for his handling of f-block elements. And the noble gases. Their atomic weights (for the most part) placed them between the halogens and the alkali metals in Mendeleev's Line. Their absolute nobility (at the time) placed them in periodic tables in a Group "0", helium above neon. Both assignments were logical, at the time, illogical, at this time.

Arguably, the Ramsay-Mendeleev classification of helium was the only reasonable one in 1900, when helium's placement above beryllium in periodic tables would have made little sense to chemists — except, perhaps, on grounds of avoiding a gap between hydrogen and helium, with its implication of missing elements, for which the atomic weights of hydrogen and helium leave, however, little room (**87**). Today, however, helium above beryllium in periodic tables yields a fuller account of the properties of the chemical elements than does helium above neon. He/Be is the simplest of –

> *Three Ways to Increase the Atomic Character*
> *of the Conventional Periodic Table*

- Place helium above beryllium.

- Elevate the footnoted "series" to block status.

- Change the block sequence from *sfdp* to *fdps*.

Because of atomic structure's arithmetical regularities, maximum atomic character for a periodic table (of atoms) means a table of maximum arithmetical regularity. It does not mean a table of maximum graphic expression of chemical relationships. Absent from periodic tables of maximum arithmetical regularity are such adjacencies as helium above neon, hydrogen above fluorine, aluminum above scandium, zinc beneath magnesium, lanthanum beneath yttrium, and uranium beneath tungsten. Historically speaking, helium-above-neon is in a distinguished company of well known, strong — if not genetically close (**83**) — chemical relationships. Just as listing elements by, e.g., atomic weight, or abundance in the earth's crust, or alphabetically requires several tables, so, too, exhibiting Periodicity's arithmetical regularities *and* its less-than-fully periodic, important, non-primary relationships requires more than one periodic table.

89. Phenomenological and Reductionist Definitions of Periodicity's Integers. Chemical thought regarding the Periodic System is enriched by having more than one periodic table, alternative sets of column labels (**94**), and complementary sets of definitions of chemical kinships. "A foolish consistency," said Emerson, "is the hobgoblin of little minds." Defining different types of chemical kinships in terms of locations of Groups within periodic tables (**79, 80**), rather than, as in today's reductionism, in terms of the electronic structure of atoms (**81**), might appear, however, on the face of it, to be inconsistent with the Periodic Table Utility Theorem: that, in the long run, the greater a Table's atomic character, the greater its usefulness. In fact, the apparent inconsistency turns out to be, in a deeper sense, the exception that confirms the wisdom of following, sometimes, in the footsteps of a contemporary of Mendeleev (1834 – 1907), Rudolph Clausius (1822 – 1866), in his establishment of the foundations of thermodynamics without reference to the atomic model of matter, deemed at the time controversial (g).

Sometimes it's useful, said Feynman, *to see how little one needs to use to arrive at a given result.* That exercise, in the case of the Periodic System, is a phenomenological, chemical, and heuristically useful account of Periodicity's leading features, illustrated in Figure 1 (and, e.g., Section **45**), meaningful to novices to chemistry, in that it does not require for an understanding of its logic a knowledge of atomic structure. Minimalism has another virtue.

There's safety in reticence. Silence may be a virtue. Because purely phenomenological accounts of phenomena do not go beyond the phenomena — and, consequently, are less explanatory, and, hence, in a sense, less satisfactory, than fundamental, reductive accounts, — because of that reserve, phenomenological accounts are less likely than reductive accounts to lead to rules that suffer from exceptions. The "Row Occupancy Rule" of Section **5**, for instance, in saying nothing more — nor less — than what is observed, admits of *no exceptions.* Its restatement as an Orbital Occupancy Rule, Section **50**, is subject, however, to a number of exceptions (**67**). For every gain (in explanatory power) there's a loss (of empirical certainty). Definitions of chemical kinships, phenomenologically and electronically (**79–81**), illustrate the same point, as do definitions of the integers e and N.

The definition of e as a Group's ordinal number in its block is unambiguous, as is the definition $N = e$ in the s-block and $N = e + 2$ in other blocks. Orderly progression of e and N across the d- and *f*-blocks is interrupted, however, if *e* is defined as the number of electrons of the same type as an atom's differentiating electron.

Similar difficulties are encountered if N is defined as the number of valence-shell electrons, or, since more than one electronic shell may be involved in a "valence shell", as the number of electrons in oxidizable subshells for the most oxidizable atoms in a particular atom's Group. For by that definition, the "natural" column numeral for the zinc group is II, not XII (**94**) — unless one argues, e.g., that zinc's $3d^{10}$ subshell is "slightly oxidized" in "back-bonding" interaction of the "soft" acid Zn^{+2} with, e.g., the "soft" ligand S^{-2} in, say, insoluble zinc sulfide; and/or unless one believes in the existence of Hg^{+3} (**44**).

Actually, the phenomenological and electronic definitions of *e* and N work in concert, by identifying, jointly, "exceptional" electron configurations, which raise the interesting question: How does one account for them? Or, failing that: How might one embed them in a broader context? (**67**), as attempted, e.g., for the zinc "anomaly" in the following section.

At issue is a critical test of the LSPT format of the Periodic System. Does it clarify the unusual character of the zinc-magnesium relation? The relation is similar, from the point of view of atomic structure, in some — and in one instance, in all — respects to each of the three types of chemical kinships, primary, secondary, and tertiary. And yet, on phenomenological grounds, it is unlike each of them in some respect.

90. The Zinc-Magnesium Test of the Left-Step Periodic Table and the Concept of *ℓ*-Nobility.
Like the laws of classical thermodynamics, systems of perfect regularity live dangerously. Any irregularity, or exception, is one irregularity, or exception, too many. The entire edifice of the LSPT as a scheme for displaying and defining chemical kinships collapses if it fails to suggest a way to account for presence of the Zinc Group in the *d*-block instead of in the *s*-block with Periodicity's other s^2 Group.

Zinc and magnesium have the same number and same type of valence-shell electrons. In a word, they are, in Jensen's terminology, "isovalent". They belong, accordingly, in the same Group, an assignment easy to display in a step-pyramid/dyad-separated periodic table, by connecting Mg and Zn by a primary kinship tie-line. Graphic representation of various kinds of chemical kinships is (to repeat) a step-pyramid table's leading virtue. It is, also, from the standpoint of block-type periodic tables, the step-pyramid tables' leading weakness: possible inclusion of primary-kinship tie-lines that cannot be represented in block-type periodic tables without awkward partitions of blocks.

Like most block-type periodic tables, the LSPT places Zn and Mg in different blocks. Is, therefore, their relation a secondary kinship? Zn and Mg's valence-shell electrons are not, accordingly, of different types. Also, although the two elements have the same maximum states of oxidation, they do not lie the same distances from their block's left ends. They lie the same distances from their blocks' right ends. Is, therefore, their relation a tertiary kinship? Zn and Mg do have non-positive electron affinities. In that sense they're "iso-acceptors". On the other hand, unlike participating atoms in the

H-F and He-Ne tertiary kinships, Zn and Mg atoms haven't different numbers and different types of valence shell electrons.

Admittedly, as in the case of tertiary kinships, Mg, Zn, and Cd do form a triad. It is of the type sdd. The first element is the odd element out. Mg is, however, unconventionally, not the first but the third element of its Group. On the other hand, 3, like 1, is an odd number.

All things considered, in summary, the Zn-Mg relation appears to be, by a small, but significant margin, a *special kinship.*

Through the designation "special kinship", the phenomenological definition of chemical kinships may seem to save the LSPT from the indignity imposed on it by a reductive, purely electronic definition of kinships, which makes the Zn-Mg relation, indisputably, a primary kinship, which is, absolutely, incompatible with the shape of the LSTP, which is, undoubtedly, the Table's chief claim to "fame".

Merely calling the Zn-Mg relation "special" does not, however, account for it. Left unaccounted for is the fact that it appears to be the LSPT's *only* special kinship. Its ad hoc designation leaves the Zn-Mg phenomenon unconnected to any other features of the Periodic System. Confidence in the System, as a *system* of interrelated phenomena, suggests seeking support for zinc's location in the *d*-block — rather than in, e.g., a bifurcated Beryllium Group (42); or in an extended *p*-block of seven members (43) — through a block-to-block trend that calls for an element at the end of the *d*-block's first row that, unlike the row's other elements, has a noble 3d subshell.

Indeed, the same sort of thing — increased nobility — occurs at the end of the first-row of the *p*-block, in neon, and, to even a greater degree, at the end of the first-row of the *s*-block, in helium. In the other direction from zinc, at the end of the first-row of the *f*-block, and in the other direction of nobility, stands ytterbium with an oxidizable $4f^{14}$ subshell. Present is a Jensen-type, block-to-block trend in the nobility of the first occurrence in Bohr's Aufbau Process of a full subshell of a given type. Professor Frank Weinhold calls the phenomenon (21) -

$$\ell\text{-}Nobility$$
$$He \quad > \quad Ne \quad > \quad Zn \quad > \quad Yb$$
$$s^2 \qquad p^6 \qquad d^{10} \qquad f^{14}$$

From the point of view of block-to-block trends, the *d*-block's terminal group, although distinctive within its block in the nobility of its outer subshell, is not distinctive in that regard *within the Periodic System as a whole.* Harbingers of zinc's noble d^{10} subshell are the ground electronic states of the coinage metals (Cu, Ag, and Au) and Pd.

Thus, by way of a block-to-block trend in ℓ-nobility, the LSPT suggests a resolution of the Zn-Mg issue. The assignment He/Be, in conjunction with the properties and locations in periodic tables of Ne and Yb, helps to rationalize zinc's chemical character vis-à-vis its location in periodic tables. Reciprocally, the trend in ℓ-nobility established by Yb, Zn, and Ne supports the assignment He/Be, Figure 38 (following page).

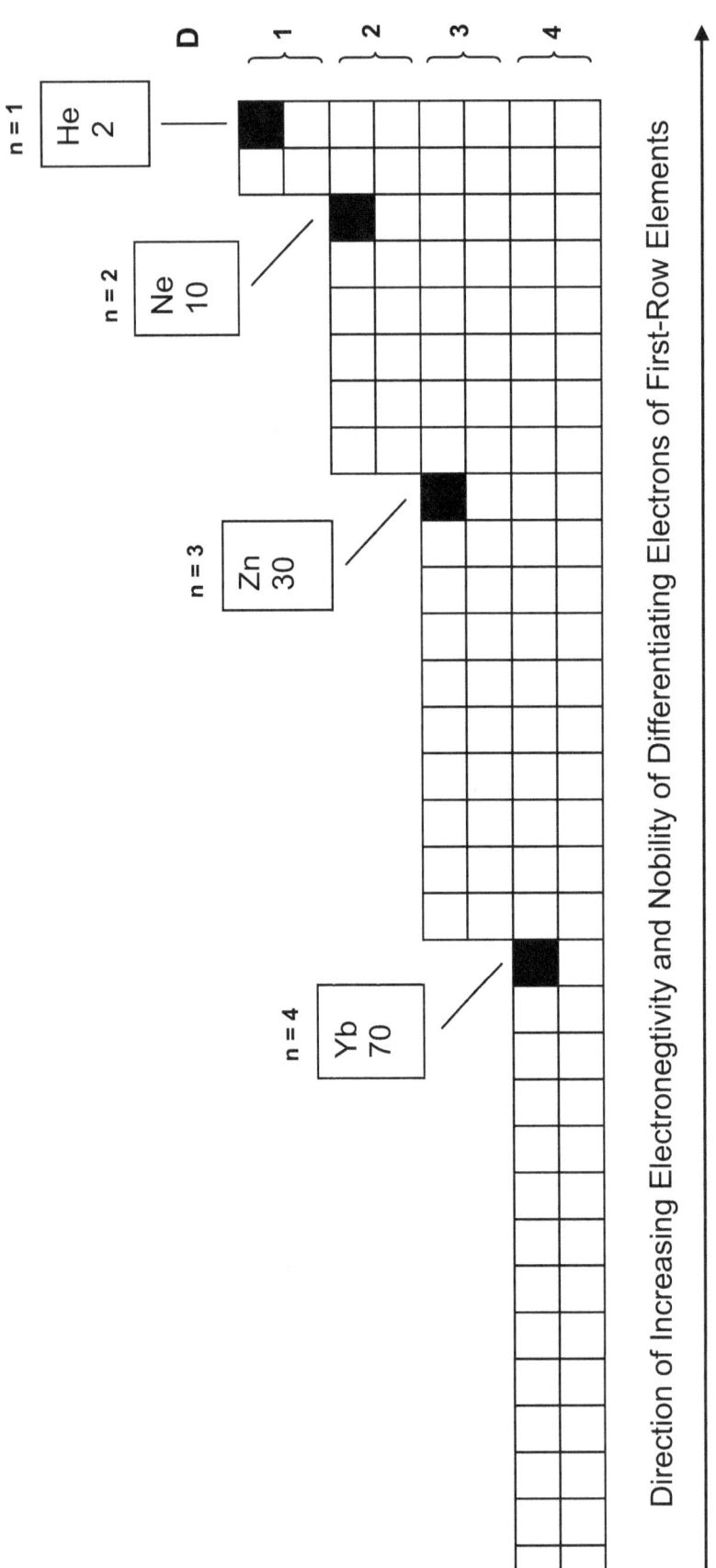

Figure 38. Location and nobility of elements where shells are first fully occupied.

As illustrated in Figure 38, the concept of ℓ-nobility and the Rule of First Full Occupancy of an Atomic Shell are complementary features of the Periodic System. Atomic shell n is first fully occupied at the end of the first row of block $\ell = n - 1$, at atomic number -

$$Z_n = 2[6n + (n\text{-}1)n(2n\text{-}1)]/3 - 2$$

Numerical values of Z_n for $n = 1$, 2, 3, and 4 are, respectively, 2 (He), 10 (Ne), 30 (Zn) and 70 (Yb). Their relative magnitudes exhibit a Jensen-like, block-to-block trend: $s <<< p << d < f$.

In summary: The initially disturbing disjunctions — that atomic spectroscopy does not classify helium as a Noble Gas; that descriptive chemistry does not classify the volatile metals as "transition elements"; and that the natural Table Construction Conventions do not place the Zinc Group among the "Main Groups" of the p-block — form, in fact, through the LSPT, a harmonious whole.

By aiding recognition of previously overlooked regularities, the shape of the LSPT helps to justify its *fdps* arrangement of blocks and its location of helium above beryllium. By promoting recognition of the phenomenon of a block-to-block trend in "ℓ-nobility", the LSPT's helps to rationalize appearance of the non-"transition metal"-like volatile metals at the end of the d-block.

Experimental support for the assertion that zinc's d^{10} subshell is "(barely) noble" (and, therefore, fits into a block-to-block trend in subshell nobility) is provided by a report of electrochemical production in an inert solvent at low temperatures of short-lived Hg(III) (44). Significance of that phenomenon regarding the issue at hand has been discounted, owing to the phenomenon's transient character, and to its electrochemical method of oxidation — for, of course, any atom can be oxidized physically all the way up (or down) to its bare core. Nowhere in the definition of oxidation states is it stated, however, what an oxidized species' half-life must be. And electrochemical oxidation, with batteries (or with electric power from a coal, oil, natural gas, or in-some-other-manner chemically powered power plant) comes down, in the end, to chemical oxidation by astutely coupled chemical changes. Be that as it may, beyond Hg(III) lies the possibility of even more stable 112(III). With respect to oxidation, zinc may stand to its (super)heavy congeners as two other elements (He and Ne), first in Groups at right ends of their blocks, stand to their heavier congeners.

91. Additional, Higher-Order, Block-to-Block Meta-Trends. "The periodic law has given the first chance of predicting not only unknown elements," remarked Mendeleev (ref. 5, p273), "but also of determining the chemical and physical properties of simple bodies still to be discovered, and those of their compounds." Similarly, by its shape, the Left-Step Periodic Table has given the first chance of identifying not only "natural" locations of "problem elements" in the Periodic System, but also of relating to each other by novel "atomanalogies" and block-to-block trends previously unrelated properties of atoms and simple substances.

All periodic tables that exhibit Groups' members in vertical columns exhibit vertical trends in metallic character. Conventional, $s(f)dp$ tables exhibit, in addition, horizontal trends in metallic character (imperfect, unless the d-block is ignored) and, also, the vector sum of those trends, in the p-block, in a diagonal trend in metalloid character. The *fdps* Periodic Table retains those vertical trends and the diagonal trend and replaces the horizontal metal-to-nonmetal trend along periods with block-to-block trends within the *fdps* Table. In the mind's eye, one can see, of course, any trend with any periodic table — but with more ease for some trends with some periodic tables than with others. Trends along conventional, *sfdp* periods are not easily seen with, e.g., Mendeleev's secondary-kinship table (Figure 23). On the other hand, secondary chemical kinships are not immediately apparent with block-type tables. And block-to-block trends have, for the most part, not been recognized at all, owing to: (i) irregular arrangements of blocks, by size, in conventional periodic tables; (ii) placement of helium above neon; and (iii) relegation of the *f*-block to a footnote. To paraphrase a remark by Mendeleev about the Periodic Law itself, in the LSPT format of blocks, many relations are observed that otherwise have escaped attention. "Nobody thought of looking for such relations [as I found] before the Periodic Law," adds Mendeleev, "which proves it to be a natural and true law."

Obvious in the format of the LSPT are block-to-block trends (*f* to *d* to *p* to *s*) in blocks' heights (2 4 6 8, for Z through 120) and blocks' widths (14 10 6 2). Ordinal numbers of blocks' first rows in the Periodic System are 1, 3, 7, 13. Relative magnitudes exhibit the usual Jensen-like, block-to-block trend: $s <<< p << d < f$. Average ordinal numbers of blocks' first-row elements are given by the expression -

$$Ave.\ Z_{block\,\ell}\ (first\ row)\ =\ 4\ell(\ell^2 + 2)/3 + 2\ell^2 + 1.5$$

For blocks 0, 1, 2, and 3, Z(average) is 1.5, 7.5, 25.5, and 63.5, as one easily verifies by computing the average of the atomic numbers of the first and last elements of blocks' first rows. The Z(average) Rule requires that helium be located in the $\ell = 0$ block and that the first row of the *f*-block start with element 57 (La) and end with element 70 (Yb). Relative magnitudes of Z(average) exhibit the usual Jensen, block-to-block trend: namely, $s <<< p << d < f$. Figure 39 below exhibits three additional block-to-block trends.

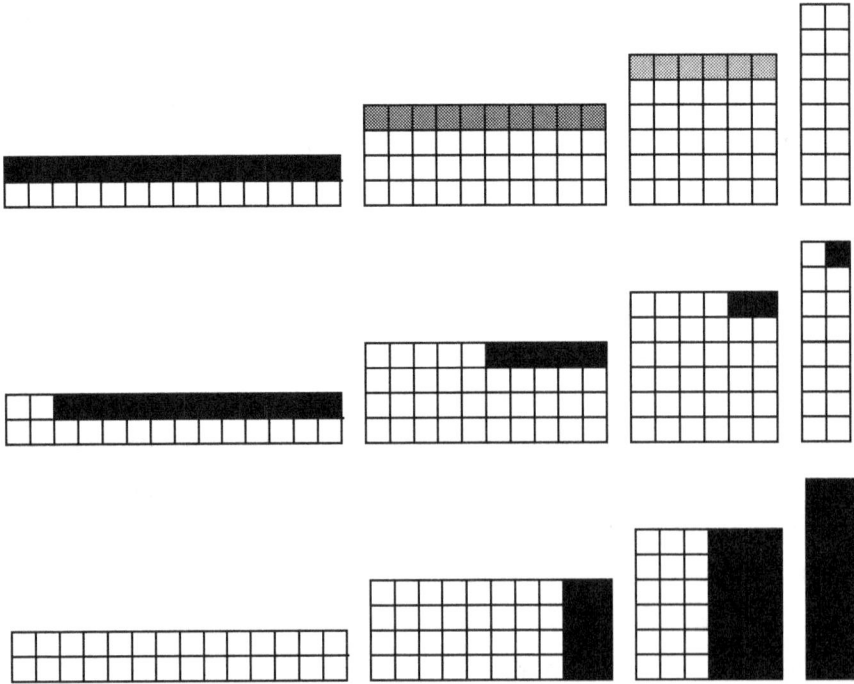

Figure 39. Three block-to-block trends in the format of the LSPT (see text).

Exhibited schematically at the top of Figure 39 is a block-to-block trend in horizontal, intra-block trends in the distinctiveness of blocks' first-row elements among themselves: extraordinary for the *f*-block's rare earths, all metals, commonest oxidation state III; well-known for the *d*-block's first-row elements, all metals, commonest oxidation state II for half its elements (Mn, Co, Ni, Cu, and Zn); much less pronounced for the first-row elements of the *p*-block, whose first element is a metalloid, whereas the other *p*-block first-row elements are nonmetals, and whose commonest oxidation numbers alternate between being even and odd; and, arguably, least of all for hydrogen and helium, which have, e.g., by far the largest difference in ionization energy of any two adjacent elements in *fdps* periods, Figure 40.

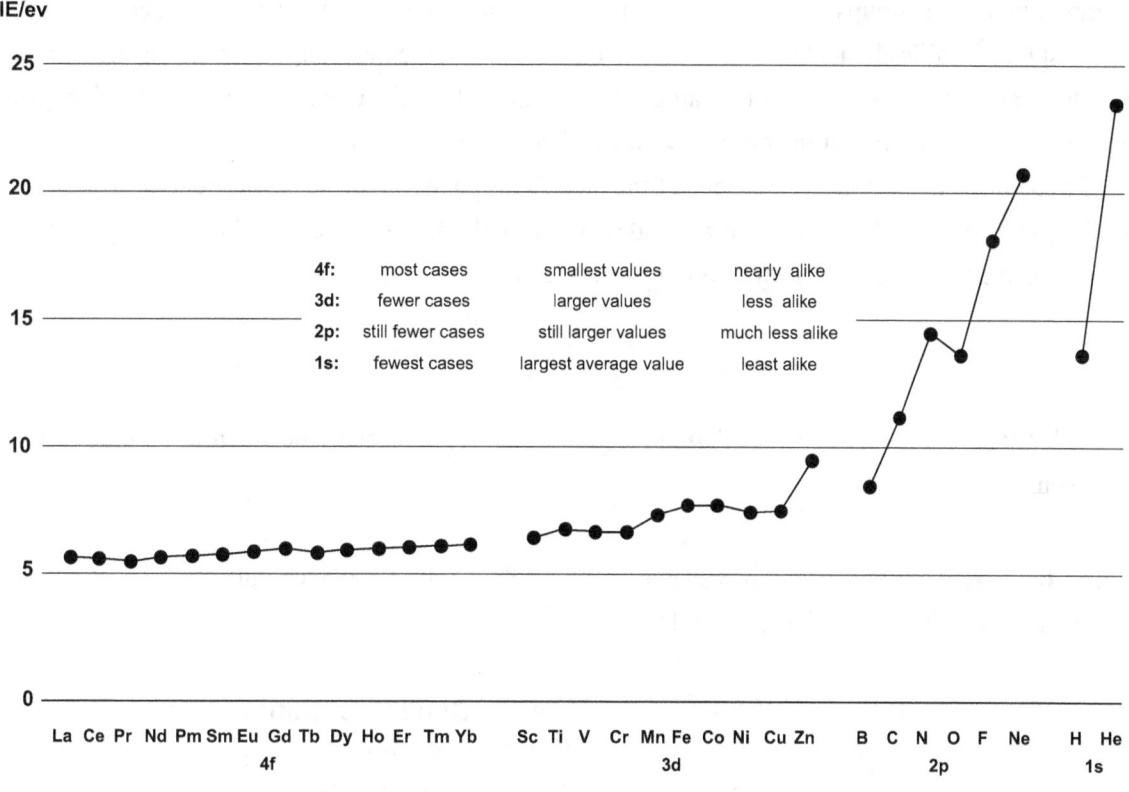

Figure 40. Atoms' first-stage ionization energies in the format of the LSPT.

Table 9. Leading numeric features of first-row elements' IEs (in eV).

	f	*d*	*p*	*s*
IE of Row's First Element	5.58	6.54	8.30	13.6
IE of Row's Last Element	6.25	9.39	21.5	24.6
Average Change in IE	0.054	0.32	2.54	11.0
Last Increment in Row's IE	0.07	1.67	4.14	11.0

Magnitudes of fifteen of the sixteen entries in Table 9 fall in the order: $s >>> p >> d > f$. Ratios of heats of vaporization of first-rows' last two elements exhibit a similar trend.

$\Delta_{vap}H$ Ratios: Tm/Yb 1.55   Cu/Zn 2.69   F/Ne 3.77   H/He 5.61.

The second diagram in Figure 39 displays a block-to-block trend in number of first-row elements whose maximum states of oxidation are less than their number of valence-shell electrons. Here the block-to-block trend is $f \ggg d \gg p > s$.

The bottom diagram in Figure 39 displays a block-to-block trend in percentage of Groups that have common names:

s-block		*p*-block		*d*-block		*f-block*
100	*>>>*	*50*	*>>*	*20*	*>*	*0*

Commonly named Groups include: the most metallic elements (the alkali and alkaline earth metals); the most nonmetallic elements (the halogens and chalcogens); the most malleable metals (the coinage metals, especially gold); the most volatile metals (especially mercury); and the most noble gases (except for helium). Present in commonly named Groups are all of Dobereiner's authentic triads.

First-row elements whose subshells of the block's preponderant type of differentiating electrons are not oxidizable are He in the *s*-block, F and Ne in the *p*-block, and Zn in the *d*-block. The percentage of such elements in the blocks' first rows falls in the order –

s	>	p	>	p	>	s
50		33		10		0

Table 10 exhibits a block-to-block trend in valence-shells' maximum occupancy, for blocks' first-row elements.

Table 10. Maximum occupancy of valence-shells of blocks' first-row elements (and those valence-shells' usual occupancy in blocks $\ell = 0, 1$).

ℓ	$(\ell + 1)$	$2(\ell + 1)^2$	**Valence-Shell Occupancy Rule**
0	1	2	Duet Rule
1	2	8	Octet Rule
2	3	18	18-Elecrtron Rule
3	4	32	32-Electron Rule

The integers 2, 8, and 18 of the Duet Rule for elements of the first-row of the *s*-block, the Octet Rule for the first row of the *p*-block, and the Eighteen Electron Rule for first row of the *d*-block are given by the expression $2(\ell + 1)^2$. To reach the 32-electron maximum given by that expression for $\ell = 3$ of the *f*-block, ions that have 8, 10, 12 and 14 *f*-electrons must have electron-pair-coordination numbers of, respectively, 12, 11, 10, and 9 from, e.g., the corresponding number of monodentate ligands. The Yb(III) ions come close to that number in YbF_3, with 13 4f electrons and a bicapped trigonal prismatic coordination of 8 fluoride ions, for a total valence-shell electron occupancy of 13 + 2 x 8 = 29. The corresponding number for a number of lanthanide compounds is 24. For Lu(III) in LuF_3 the count is 30. That's not a violation of the d-block's 18-electron rule, inasmuch as Lu is not a first-row element.

The frequency with which the *maximum* occupancy for first row elements' valence-shells is their *usual* occupancy decreases from block-to-block in the order: $s > p > d > f$ (h).

Another block-to-block trend through the p-, d-, and f-blocks is the number of Groups per block of hypovalent elements: i.e., elements that are electron-deficient with respect to formation of conventional, electron-pair, two-center bonds. Hypovalency occurs, accordingly, when the number of valence-shell electrons is less than the number of valence-shell orbitals. Since the latter is $(2\ell + 1)$ plus 1 (for an always available s-orbital), the number of Groups of electron deficient elements in block ℓ (> 0) is $[(2\ell + 1) + 1] - 2$(for electrons from an outer s-subshell) $- 1$(for electron-deficiency) $= 2\ell - 1 = 1, 3, 5$ for $\ell = 1, 2, 3$. The p-block's sole hypovalent Group is the boron Group. Blocks' first non-hypovalent Groups are ones whose atoms have half-filled valence-shells: e.g., C and Cr.

Table 11 (following page) exhibits extrapolations from the f-, d-, and p-blocks to a two-column block capped by two low-boiling, small-core, large-ionization-energy nonmetals, one reactive, one unreactive (such as H and He).

Most of the block-to-bock trends cited in this section and elsewhere in this report are anchored, at the outset, or capped off, at the end, by He/Be. The trends have been commonly overlooked, as mentioned, for three reasons: placement (understandably, in the early history of the Periodic System) of helium above neon; placement (for chemical reasons) of s-elements on the left; and placement (historically) of the lanthanides and actinides (such as were known) (i) within the d-block, initially, and, later, (ii) in a super-Group III, and usually, today, (iii) beneath a (consequently) truncated periodic table, in two special "series".

ADDED NOTE: **Brief History of the Roots of the Rule of First-Element Distinctiveness and Its Use in Support of Location in Periodic Tables of Helium above Beryllium.** Seven episodes in the evolution of the idea of using the Mendeleev-Jensen Rule in support of the LSPT and its location of helium above beryllium — with all the consequences pertaining thereto — stand out in the author's mind. First was Mendeleev statement of the Rule of Distinctiveness of *Light Elements* (**30**). Subsequently, textbooks of inorganic chemistry emphasized the distinctiveness with respect to their congeners of Nature's two lightest metals: *lithium* and *beryllium*. In 1977 the Russian chemist Shchukarev stated an extension of Mendeleev's Rule to *Groups' first-row elements* (**32**). In 1986 Jensen went still further, in a statement of a *block-to-block trend in distinctiveness of Groups' first-row elements* (**31**). Jensen's extension of the Mendeleev-Shchukarev Rule seemed to have passed unnoticed, except in one instance. S*pecific instance*s of the MSJ phenomenon began to accumulate in the present author's files around 1990. In 2000 most of the trends were included in a manuscript, solicited by an editor, Eric Scerri, but never published (under the present author's name, by said editor), titled, in its final, 83-single-space-page version (which contains a significant part of this report) "*f d p* s". There, for the first time, in print, the MSJ Rule was used in *support of location of helium*

Table 11. Extrapolations from the *f*-block through the *d*- and *p*-blocks to the *s*-block, predicted to contain for its first-row two low-atomic number, low-boiling, nonmetallic elements of small core radii, one reactive, one not.

Properties of First-Row Element(s)	Observed f-Block Lanthanons	Observed d-Block Scandons	Observed p-Block Borons	Extrapolations s-Block x	Extrapolations s-Block y	Observed s-Block H	Observed s-Block He
Number of Elements	2 x 7	2 x 5	2 x 3	2 x 1		2	
Z of First Element	57	21	5	1*		1	
Z of Last Element	70	30	10	1	+ 1		2
Physical States, STP	s(olid)	s	s s g g g(as)	g	g	g	g
nbp/K	hi	hi	hi hi 77 90 85 27	<77	<27	20	4
First Element's IE/eV	5.6	6.5	8.3	>8.3		13.6	
Last Element's IE/eV	6.2	9.4	21.6		>21.6		24.6
Metallic Character	very hi	hi	metalloid/nonmetals	nonmetals		nonmetals	
First core radius r/Å	1.16 (La^{+3})	0.87 (Sc^{+3})	0.2 (B^{+3})	<0.2		0	
Last core radius r/Å	---	---	0.05 (Ne^{+8})		<0.05		0
Reactive? (Y Yes, N No)	Y	Y	Y Y Y Y Y N	Y	N	Y	N

$$
\begin{aligned}
*\ 57 &= (2+2) + (8+8) + (18+18) + 1 \\
21 &= (2+2) + (8+8) + 1 \\
5 &= (2+2) + 1 \\
1 &= 1
\end{aligned}
$$

above beryllium in Periodic Tables and, hence, of use of the Left-Step Periodic Table in searches for block-to-block trends and other regularities in the Periodic System. Following a public presentation in 2005 of this report's leading ideas (a), Weinhold offered a *quantum mechanical explanation* for the First-Element Rule (21).

92. The Lanthanide-Actinide Question. Whether or not the Conventional Periodic Table is, in fact, truncated (**91**) when the lanthanides and actinides are footnoted, is a moot question. Is the footnote merely a convenience for printers? Or has it scientific significance? Seaborg's actinide hypothesis (47) is not equivalent to saying that the lanthanide and actinide "series" constitute the first two rows of an "*f*-block". Not uncommon, even today, is the traditional view (48) that those fourteen — sometimes fifteen — pairs of elements are members of a special, "continuous" or "coherent" group (49).

> "The lanthanides comprise the largest natural-occurring group in the periodic table" (50).

> "The 28 lanthanides and actinides together make up about one-quarter of the whole periodic table as it is now known, with 109+ elements. Together with the four related elements Sc, Y, La, and Ac, *they all belong to transition Group III[d],* which is thus the largest subgroup in the periodic table having 32 members and comprising almost 30% of the elements" (51, emphasis added).

Consequently, the lanthanide-actinide pairs are not labeled in IUPAC's scheme of column labels. A U.S. nomenclature committee's precursor to IUPAC's scheme placed those pairs in a special group "3f" (52). Often the two footnoted "series" are displayed with a non-block-like space between them. Some schemes start with La and Ac and end with Lu and Lr (15 elements per series). Others schemes start with Ce and Th. Mendeleev considered the rare earths to be the leading unsolved problem for the Periodic Classification of the Elements (53). For most of his life the present author was completely confused as to the exact meaning of the lanthanide-actinide footnote. Most chemists he's spoken to about it have admitted to the same confusion. [cf. Jensen's parenthetical remark in note (k)]. Which of the elements, if any, are congeners of Sc and Y? One might suppose that the "long form" of the Conventional Periodic Table would answer that question, yet the table's irregularities are such that even a leading authority on the Periodic System placed not so long ago Sc and Y above La and Ac (15).

One way to handle the lanthanide-actinide issue with the Conventional Periodic Table, *f*-elements footnoted, is to place Lu and Lr beneath Sc and Y and to leave a space between the s- and d-blocks, to indicate a gap in atomic numbers after $_{56}$Ba and $_{88}$Ra in going to $_{71}$Lu and $_{103}$No, filled by the 28 footnoted elements, starting with $_{57}$La and $_{89}$Ac and ending with $_{70}$Yb and $_{102}$No, Figure 37 (k).

93. Exceptional "Series" Becoming the Rule. Transformation of the lanthanides and actinides from their exceptional positions beneath periodic tables into an important step of the Left-Step Periodic Table is similar, in its significance for Periodic System Systematics, to Einstein's transformation of the leading exception to the Law of Dulong and Petit (diamond) into the leading example of the quantum-statistical model of specific heats. Einstein's interpretation of the "diamond exception" made quantum theory popular with physicists. Perhaps the LSPT's four hundred percent increase in display of Periodicity's dyadic character, over that of the Conventional Periodic Table (which exhibits a single, split dyad: its second and third periods), through, in part, transformation of two exceptional, footnoted "series" into Periodicity's longest dyad, will help to popularize among chemists appearance of helium at the end of the first period of Periodicity's shortest dyad.

Removal of the lanthanides and actinides from an exceptional, super-Group III eliminates leading exceptions to the rule that each element has a unique location in periodic tables. In the LSPT format, the lanthanide-actinide "series" literally point through the *d*- and *p*-blocks to the *s*-block and the need for locating at its top Nature's two lightest elements (Table 11).

That location of 28 metals at the bottom of periodic tables might influence location of two nonmetals at their tops reflects the fact, frequently emphasized in this report, that the Periodic System is a *system*. Important features of the System not exhibited by the Left-Step Table include: First-Element Distinctiveness (unless the LSPT is annotated in the manner of Figure 15); dramatic changes in properties with increasing atomic number after each Noble Gas (unless the LSPT is annotated as in Figure 8); and graphic exhibition of secondary kinships. Historically, secondary kinships have been indicated in periodic tables that, like the LSPT, and the Conventional Table, have no column gaps, by alphanumeric column labels (**94**).

94. "Natural" Column Labels and "True" Periodic Tables. Periodic tables may be viewed as *models* of the Periodic Law. Some philosophical remarks regarding models (54) apply, accordingly, to periodic tables. Useful models introduce useful terminology. From the valence-bond model of molecules come the terms single, double, and triple bonds. From tabular expressions of the Periodic Law come the terms periods, groups, and blocks. Display of Periodicity's Groups is the raison d'etre of periodic tables, display of secondary kinships the chief reason for existence of step-pyramid and "short-form" tables and, as mentioned, alphanumeric column labels. A significant feature of the LSPT is its suggestion of *natural* column labels, Figure 41 (following page).

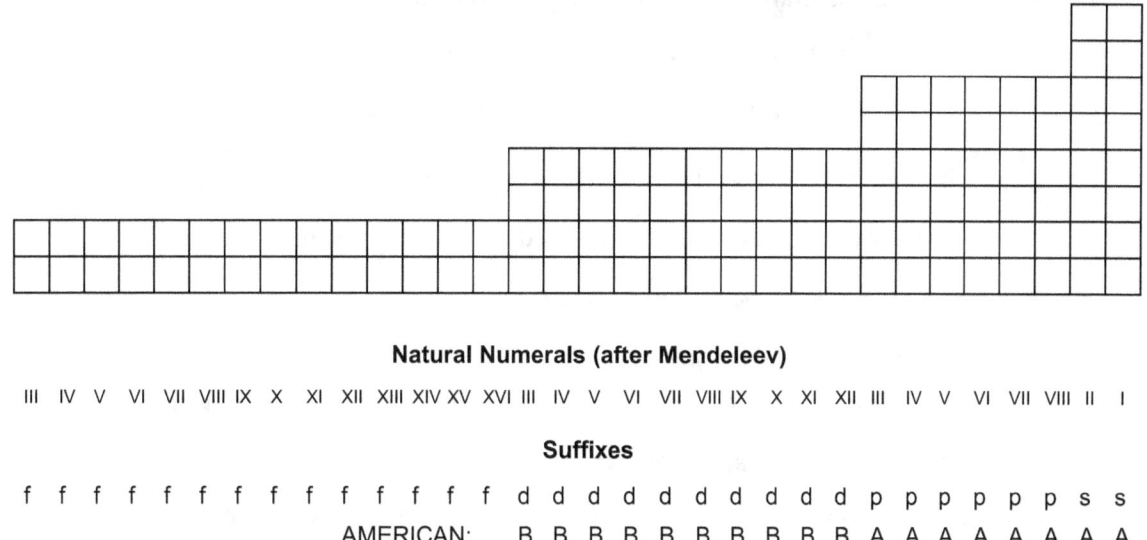

Natural Numerals (after Mendeleev)

III IV V VI VII VIII IX X XI XII XIII XIV XV XVI III IV V VI VII VIII IX X XI XII III IV V VI VII VIII II I

Suffixes

f f f f f f f f f f f f f f d d d d d d d d d d p p p p p p s s

AMERICAN: B B B B B B B B B A A A A A A A A

EUROPEAN: A A A A A A A A B B B B B B B B A A

IUPAC

3 4 5 6 7 8 9 10 11 12 13 14 15 16 17 18 1 2

Figure 41. Four column-labeling schemes.

For every gain, it's said, there's a loss. Gained when IUPAC rejected traditional alphanumeric column labels for the 1-18 scheme was brevity; distinctive labels for the iron, cobalt, and nickel groups; avoidance of an international controversy regarding use of "A" and "B" in column labels; and simplicity. A child could assign IUPAC's labels 1-18 to the Conventional, "18 Column", Periodic Table. Lost was expression of secondary chemical kinships and, for the *p*-block, statement of maximum states of oxidation, numbers of valence-shell electrons, and charges on atomic cores. For inorganic chemists and computer programmers at abstracting services, those loses are, for different reasons, unimportant. Professional chemists, unlike beginning students, needn't be reminded of the chemical information encoded in alphanumeric column labels. And computer programmers haven't any need for it (j).

The question Which column-labeling scheme is best? is like the question Which periodic table is best? (**99**). Best for what purpose(s)? Teaching? Reporting research results? Programming computers? Chemistry benefits from having several useful column-labeling schemes, some informative, some succinct, provided, of course, that that diversity doesn't create confusion in classrooms and the professional literature. No problem. Beginning students seldom read the professional literature. And few professional chemists have difficulty understanding classroom column labels, particularly since they were themselves once students. Table 12 (following page) lists virtues of the "natural" column labels suggested by the LSTP.

Table 12. Virtues of "natural" column labels suggested by the LSPT (Figure 41). They -

- indicate *all* secondary kinships (without use of the controversial suffixes "A" and "B")

- yield numbers *and* (preponderant) *types* of valence-shell differentiating electrons

- distinguish between *s*- and *p*-block "Main Groups"

- label columns of the *f*-block, unlabeled by the other schemes cited in Figure 41

- provide labels for new blocks of elements

- acknowledge, via the suffixes *s, p, d,* and *f,* that the periodic table now finds its "natural interpretation in the detailed electronic structure of the atom" (1)

- draw on the best features of previous labels

  o Mendeleev's numerals based on maximum states of oxidation

  o Suffixes, as in the European AB and American ABA schemes, to indicate secondary kinships

  o IUPAC's column numbers greater than 8 (in the *d*-block), so that each Group of a block has a distinctive number

- can be easily modified to yield a complementary set of numerals, *e*, equal to the occupancy of orbitals associated with a block's preponderant type of differentiating electrons

- are applicable to *all* periodic tables

The natural, *s/p/d/f*-suffixed column labels, unlike the older, artificial AB and ABA alphanumeric schemes, express explicitly the need to move helium from its conventional position in Group VIIIp to the top of Group II(s).

> For the *s*-block's Groups, I and II, the suffix *s*, as an aid in distinguishing Groups in different blocks from each other in schemes that indicates secondary kinships, is unnecessary, since, in the cited, natural alphanumeric scheme considered in Table 12, no other blocks have a Group numeral I or II. In, however, a complementary, perfect-regularity-maintaining, alphanumeric **e𝓵** scheme of column labels (see below), the *s*-suffix is necessary, since each block's first Group is a Group I.

Many sets of alphanumeric column labels exist. The "natural" labels referred to above might be designated N𝓵 labels, where N = number of "valence shell" electrons = number of electrons in oxidizable sub-shells of the most oxidizable atoms of Groups. A related set of labels might be designated *e𝓵* labels, where *e* = columns' ordinal numbers within their blocks (= electron-occupancy of orbitals occupied by atoms' differentiating electrons, in the s- and p-blocks, and in the majority of the cases in the *d*- and *f*-blocks). As already mentioned, **N** = *e* in the *s*-block and *e* + 2 in other blocks. The two options are displayed in the format of the Conventional Periodic Table in Figure 42 (following page).

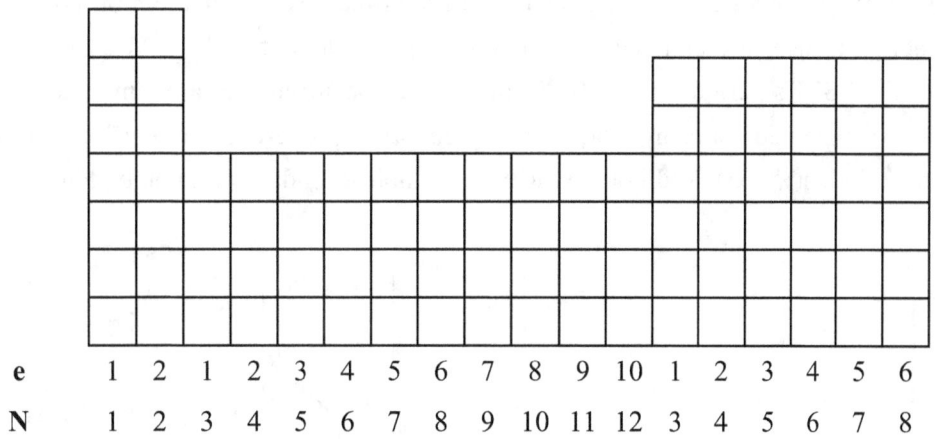

e	1	2	1	2	3	4	5	6	7	8	9	10	1	2	3	4	5	6
N	1	2	3	4	5	6	7	8	9	10	11	12	3	4	5	6	7	8

Column numbers or numerals N are referred to above as "natural" in the sense that they are generated by elements' maximum states of oxidation. They are equal to numbers of valence-shell electrons; i.e., to the number of electrons in chemically oxidizable shells or subshells. Applied to *sfdp* periodic tables, they look "natural", in that periods begins with a column labeled "1" (or "I"). Applied to the LSPT, however, they look "unnatural", in that periods, except for the first two, begin with a column labeled "3" (or "III"). A more natural set of column numbers, from the standpoint of the LSPT, are the numbers e = columns' ordinal numbers within their blocks (= occupancy of atoms' outermost orbitals in the s- and p-blocks, and in the majority of instances in the d- and f-blocks). In the e-scheme, each period in the *fdps* table begins with a column labeled "1".

In summary: To begin periods of *sfdp* periodic tables with a column labeled "1", use N-values. For the *fdps* table, use e-values.

A Traditional View of sfdp/N-labeled and fdps/e-labeled Periodic Tables

"A true periodic table," writes a well-informed correspondent, "is based on the total valence manifold occupancy of an atom [N] and not just the occupancy of the outermost differentiating orbital level [e]. Hence, I would argue that what you are really describing [in your e-labeled left-step table] is a highly articulated electron configuration filling chart and not a true periodic table. A true periodic table and a configuration chart emphasize different aspects of atomic structure. They nicely complement one another, and there is no doubt that configuration charts are implicitly periodic, just as [true] periodic tables implicitly contain configurations, but in practice they should be carefully distinguished from one another."

In the correspondent's terminology, a *"true periodic table"* is an *N-labeled*, **sfdp** table. A *"configuration chart"* is an *e-labeled*, **fdps** table. Both sets of Group labels, those involving N and those involving e, are useful when contemplating the chemistries of d- and f-block elements.

Number of d- and f-electrons on an Ion of Charge Q

$$e - (Q - 2) = N - Q$$

95. Secondary Periodicity. A noteworthy feature of the LSPT is creation of, not only natural column labels, but, also, as mentioned, a second set of period numbers, P_{fdps} ($= P_{LSPT}$). Those ordinal numbers of the LSPT's periods are useful in displaying "occurrence of a nonmonotonic course of the quantitative expression of many fundamentally important properties of atoms", sometimes called "secondary periodicity" (55). Radii of atomic cores exhibit secondary periodicity, Figure 43.

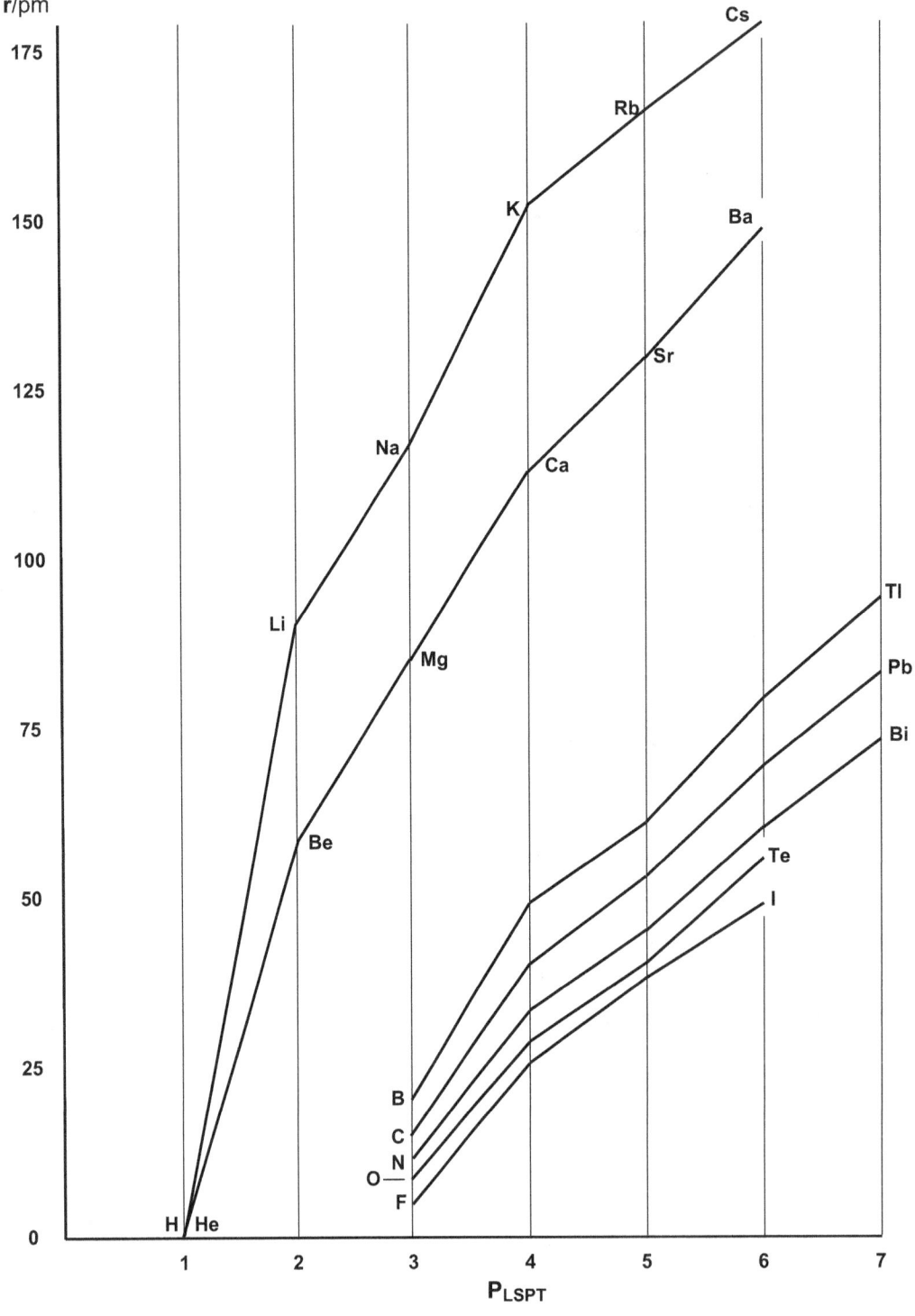

Figure 43. Plots of radii of atoms' cores vs. P_{fdps}.

From the standpoint of the overall pattern of the figure, Pauling's radius for I^{+7} seems too small, by about 5 pm. Another atomic property that exhibits a "nonmonotonic course" in rates of change within Groups is electronegativity.

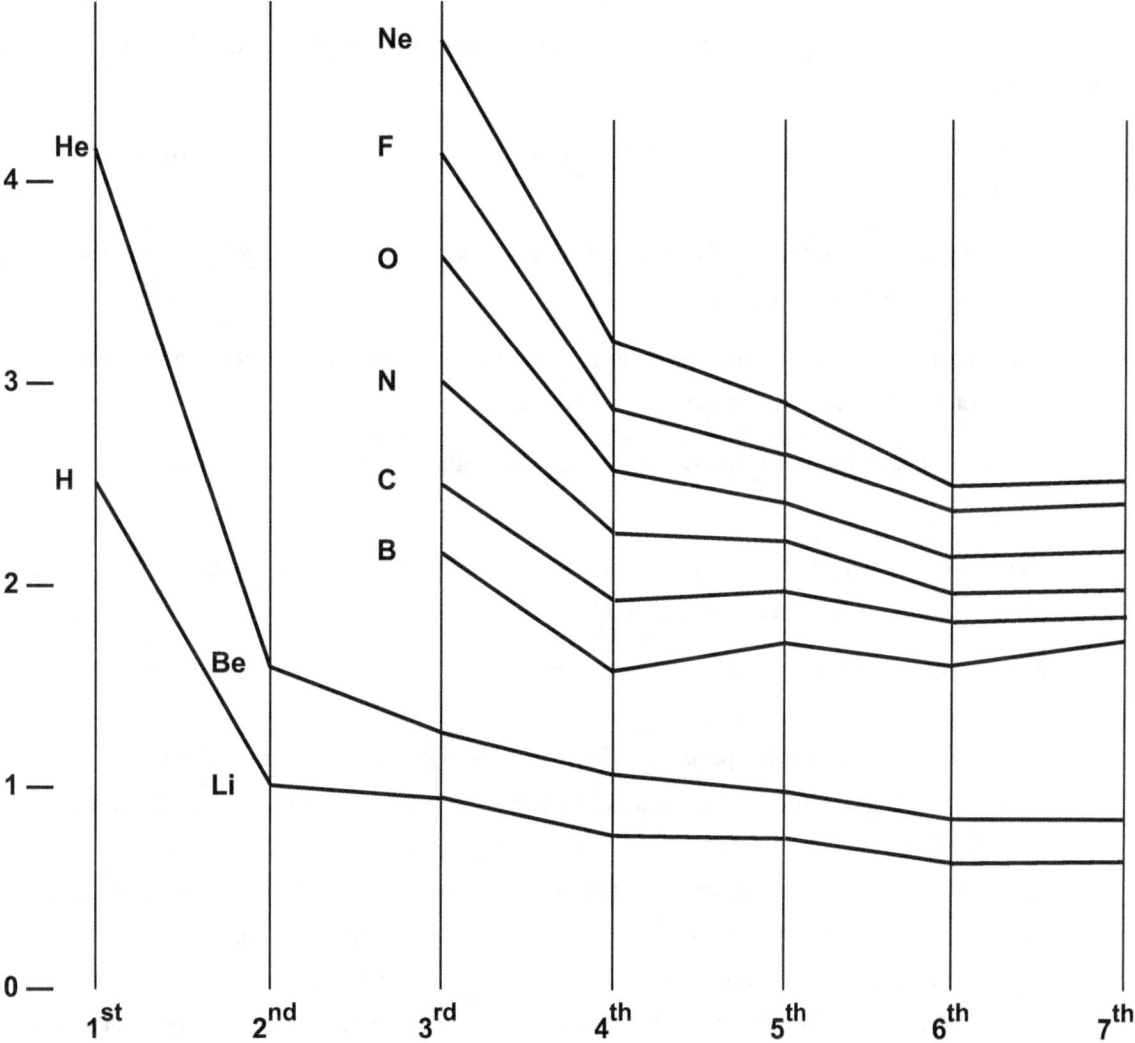

Figure 44. Plots of Allen's electronegativities (56) vs. P_{fdps}.

The overall pattern of the electronegativity plot in Figure 44 strongly suggests that the natural tie-line from helium to another element is (as shown) to beryllium and not to neon; and similarly for hydrogen: its natural tie line is to lithium, not to fluorine (or carbon). Illustrated is: (i) Jensen's observation that first-element distinctiveness (FED) is greater in the *s*-block than in the *p*-block; (ii) the suggestion (**41**) that FED increases within blocks, from left to right, from boron to fluorine and from hydrogen to helium; and (iii) congruence of secondary periodicity's kinks, as in Figure 43, provided the abscissa is P_{fdps}, not P_{sfdp}. Congruence of secondary periodicity's kinks in the *s*-block with kinks in the *p*-block (Figures 43 and 44), requires use of P_{fdps}, rather than P_{sfdp}. In the latter instance, points for the *s*-block stay put but those of the *p*-block move one unit to the left (**55**).

The significance for The Hydrogen-Helium Problem-Element Problem of the plot of Allen's electronegativities, taken in conjunction with the Mendeleev-Jensen Rule of First-Element Distinctiveness, is spelled out in Table 13.

Table 13. Remarks regarding plots of Allen's Electronegativities vs. P_{fdps}: additional reasons for He/Be and H/Li.

- The Noble Gases have a satisfactory first-element in neon, as do the Alkali Metals in hydrogen.

- Helium is not a satisfactory first-element for the Noble Gases, nor is hydrogen a satisfactory first-element for the Halogens.

- Beryllium is not a satisfactory first-element for the Alkaline Earth Metals, nor is lithium a satisfactory first-element for the Alkali Metals.

- Helium is a satisfactory first-element for the Alkaline Earth Metals, as is hydrogen for the Alkali Metals.

- Helium-above-beryllium and hydrogen-above-lithium makes First-Element Distinctiveness greater in the *s*-block than in the *p*-block.

- First-Element Distinctiveness increases from left-to-right in the *s*- and *p*-blocks.

The phenomenon of secondary periodicity provides, in summary, several additional reasons for placing helium at the head of the column of alkaline earth metals. That location for helium, to repeat, does not imply, of course, owing to the phenomenon of light-element distinctiveness, that helium is a metal, anymore than grouping carbon with the metals tin and lead implies that carbon is a metal. The first four Groups of the *p*-block are, indeed, famous for beginning with nonmetals and ending with metals. That that transition toward metallic character within Groups is striking in magnitude and suddenness in the *s*-block is consistent with a block-to-block trend in the magnitude of First-Element Distinctiveness and a block-to-block trend in onset of metallic character within groups: sudden in the *s*-block; less sudden in the *p*-block; and absent in the *d*- and *f*-blocks, because all elements in those blocks are metals.

96. Periodic Tables' Mathematically Curious Coordinate. From a mathematical point of view, periodic tables labeled with alphanumeric column labels have a horizontal coordinate that exhibits two curious features: (i) numbers (or numerals) *repeat* themselves and have (ii) attached *suffixes*. The explanation for those mathematical oddities lies, chemically, in the phenomenon of secondary kinships and, physically, in existence of different types of subshells that, for a given shell, are not, with increasing atomic number, occupied consecutively, one immediately after the other one. Created, correspondingly, are chemical "oddities", in the form of some 700 periodic tables that feature singly and in various combinations circles, spirals, helices, cones, lemniscates, squares, triangles, chess boards, zigzags, special flaps, and other space curves and geometrical constructions in symmetrical and unsymmetrical forms with interrupted and uninterrupted series (3), all because of the phenomenon of secondary kinships, owing to occurrence of orbitals that have the same principal quantum number but different angular momentum quantum numbers and, consequently, different degrees of penetration of inner-electron density in many-electron atoms and, hence, different orbital energies, and, therefore, significantly different orders of occupancy in Bohr's Aufbau Process.

Periodic tables' alphanumerically-labeled coordinates are, in other words, projections onto one dimension of two dimensions of Chemical Periodicity. One dimension, indicated by a number or numeral, is the number of valence-shell electrons, N; or the number of differentiating-type electrons e. The other dimension, indicated by a suffix, $\ell (= s, p, d, f)$, is the preponderant type of differentiating electrons.

97. "Color"-numeric Column Labels. Shown in Figure 45 (in black and white, to keep down the cost of this book) is a disentanglement of alphanumeric column labels' two components, N and ℓ, via use of "color"-numeric labels.

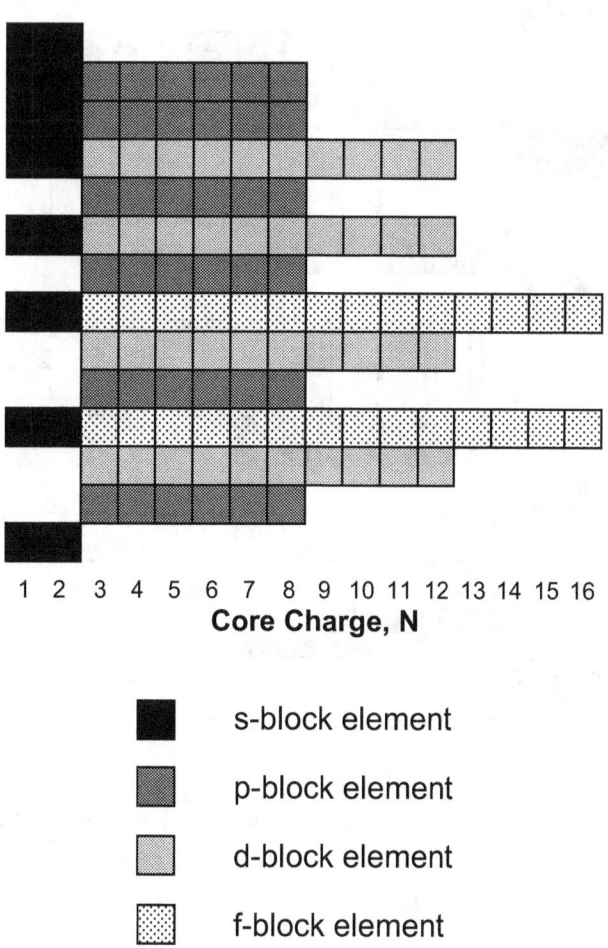

Figure 45. A Newlands-type "color"-numerically labeled periodic table.

Squares' fills in Figure 45 indicate ℓ-values. The horizontal coordinate indicates N-values. The physical significance of the table's vertical coordinate is not easily described. Passage to Figure 45 from *sfdp* and *fdps* tables simplifies the tables' horizontal coordinate at the expense of simplicities of interpretation of the vertical coordinate. In Figure 45 all elements that have a particular maximum oxidation number, e.g., +3, appear in the same column. Sans squares' fills, Figure 45 is a Newlands-type display — a truly "short form" table, in that it has only 16 columns, not the usual 32.

Assigning, in order, to each of the 20 orbital types exhibited in Figure 45 a distinctive vertical coordinate, namely R (the Row-Orbital Ordinal Numbers in the Periodic System), produces Figure 46.

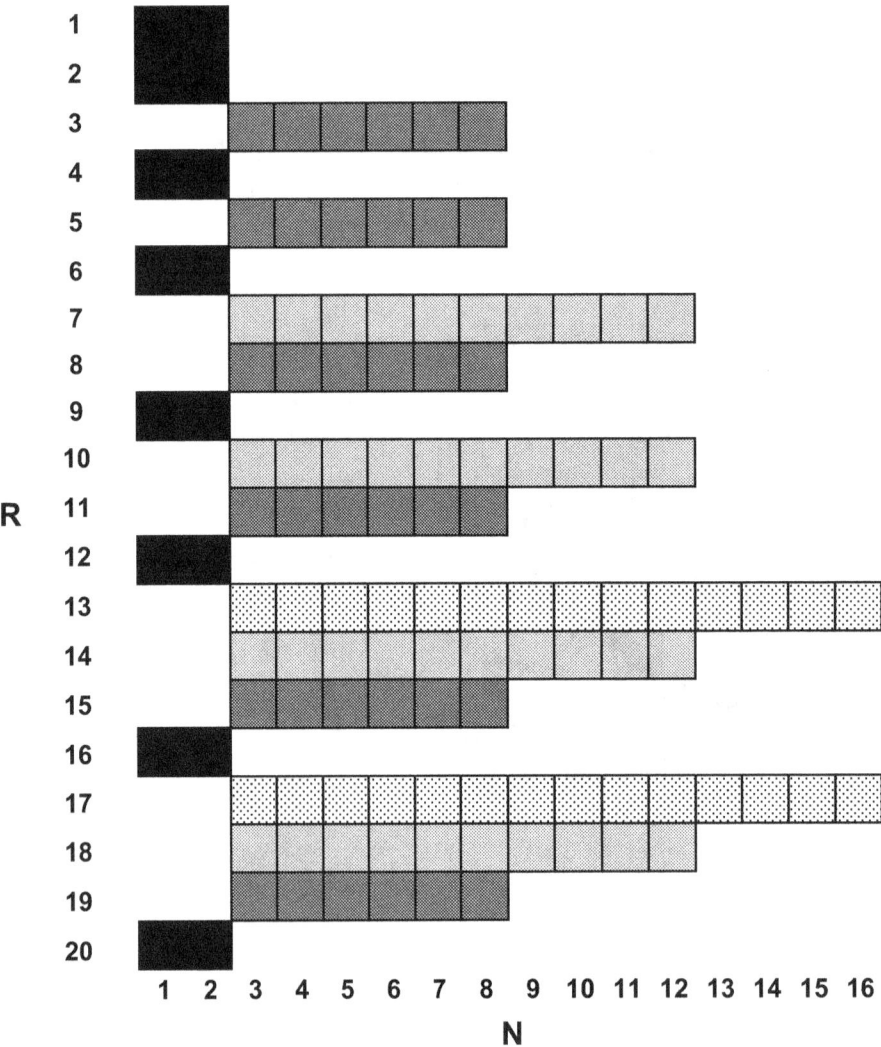

Figure 46. A Newlands-type table whose vertical coordinate is rows' ordinal numbers in the Periodic System.

Figure 46 in true color appears on a page facing the back cover of Mazur's book (3). The number of columns in it is further reduced, to 14 (for Z through 120), if, instead of plotting horizontally core charge N (= total number of valence-shell electrons), one plots *e*: number of electrons in the subshells that are being filled, with increasing Z (= Groups' ordinal numbers within their blocks), Figure 47.

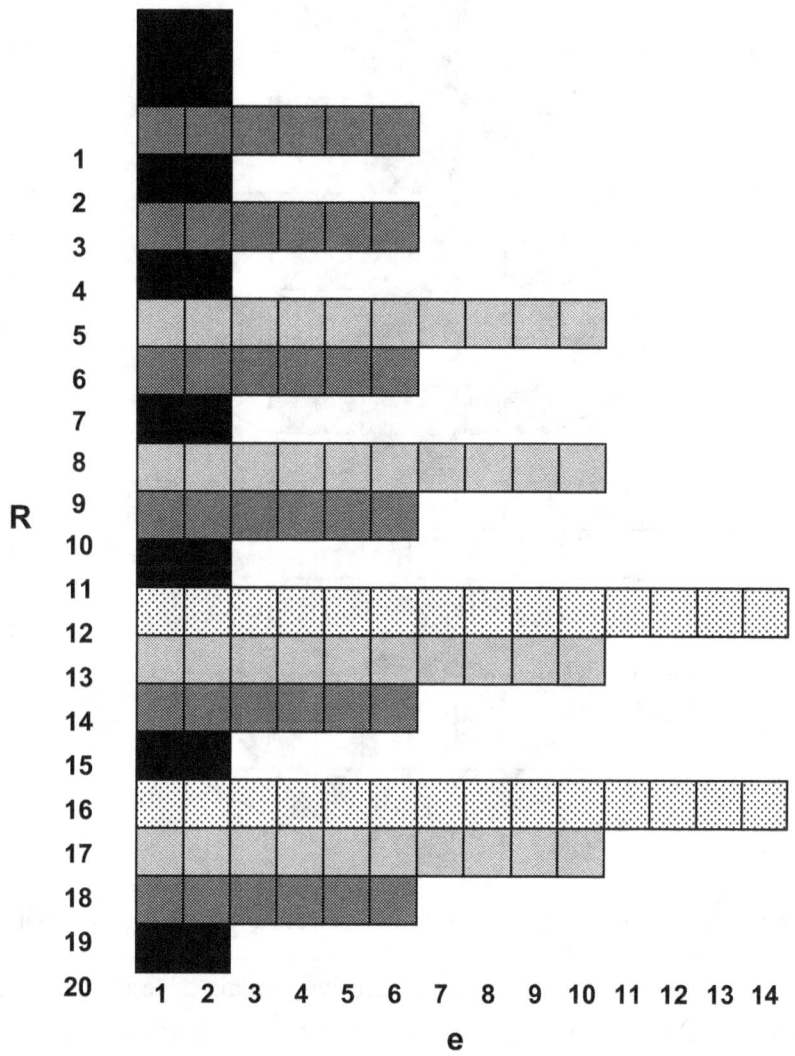

Figure 47. A "color"-numerically labeled R/*e* periodic table.

Centering rows of Figure 47 yields Figure 48.

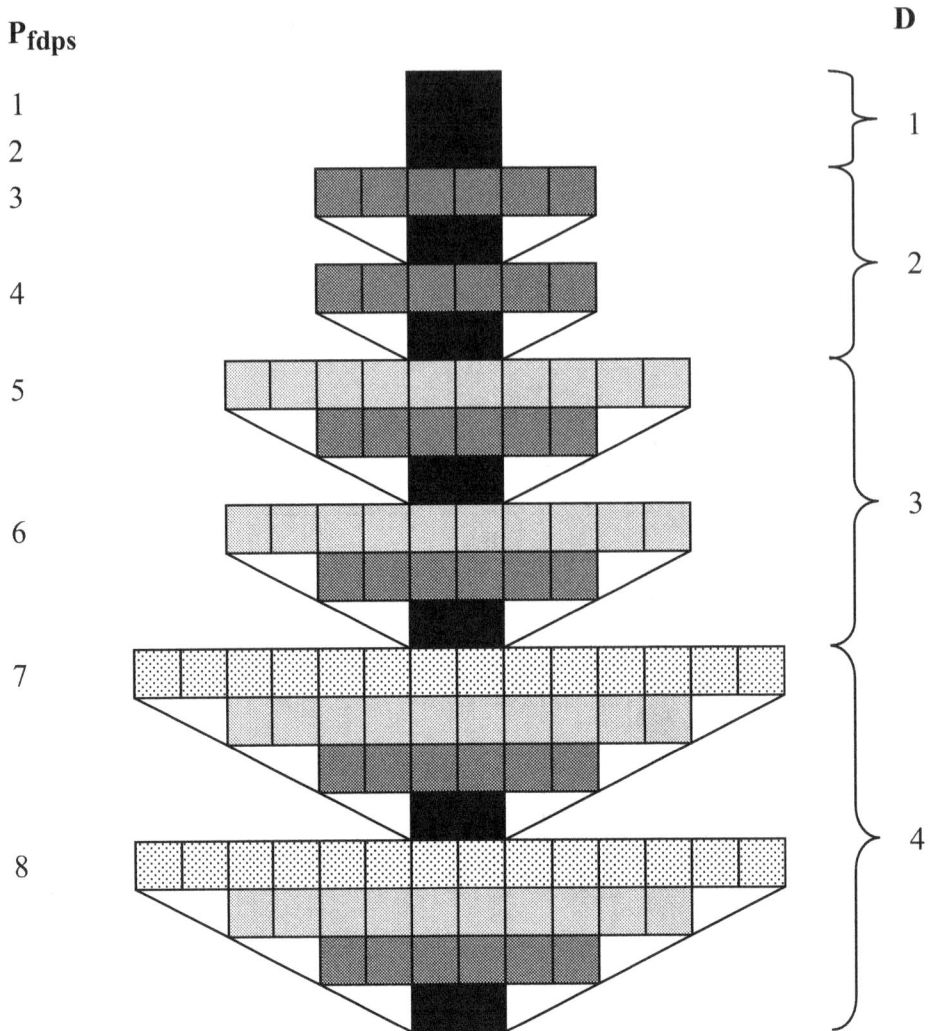

Figure 48. A bilaterally symmetric "color"-numerically labeled periodic table.

Figure 48 is essentially identical to Mazurs' favorite periodic table (57). It reveals, as well as any periodic table does, Periodicity's arithmetical regularities (for Z ending at, e.g., 120). And by its location in Mazurs' book (on the inside of its front cover), it reveals mankind's love of regularity, and symmetry, achieved in Figure 48 at the expense, however, of chemical and physical significance for the table's horizontal coordinate. That is true of all step-pyramid type periodic tables.

Figure 49 (following page) may be viewed as a projection onto two dimensions of a vertical compression of Figure 48, jointly with a rearrangement into a third dimension of *rows* of the *p*-, *d*-, and *f*-blocks into *concentric, horizontal rings* about a central *s-block circle.*

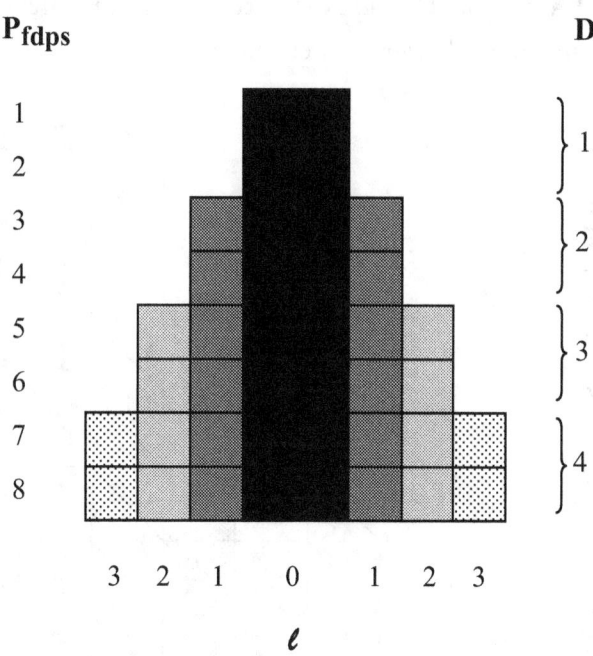

Figure 49. Side-view of the three-dimensional "Periodic Round Table".

Figure 49 is, in other words, a front view of a vertical plane through a three-dimensional arrangement of the elements in which elements of, e.g., the seventh *fdps* period appear on a horizontal disk whose outer ring contains the seventh period's fourteen 4f-elements about a ring of its ten 5p-elements about a ring of its six 6p-elements about a central circle of its two 7s-elements. Created when the elements' symbols are inscribed on wood disks is a "Periodic Round Table" (available in an attractive construction from Gary Katz, P. O. Box 156, Cabot, VT 05647, tel. (802) 563-2078). It is three dimensional arrangement of atoms arranged along cylindrical-polar coordinates. Its regularity arises because (i) the circumference of a circle is a linear function of its radius (set equal to ℓ in the PRT) and because (ii) the number of Groups for a given value of ℓ is a linear function of ℓ. Visible in the assembled table are symbols of all of the Periodic System's first-row elements, and only those elements. Visible on removal of, e.g., the top disk (the H-He disk) are Li and Be.

[Katz discusses his PRT and the LSPT in an article titled "The Periodic Table: An Eight Period Table for the 21st Centrury ," *Chemical Educator* **6**, 324–332 (2001).]

98. Arrangements of Atoms in PℓN and Pℓe Space. Shown in Figure 50 is a disentanglement of N and ℓ via use of *rectangular coordinates* in three dimensions.

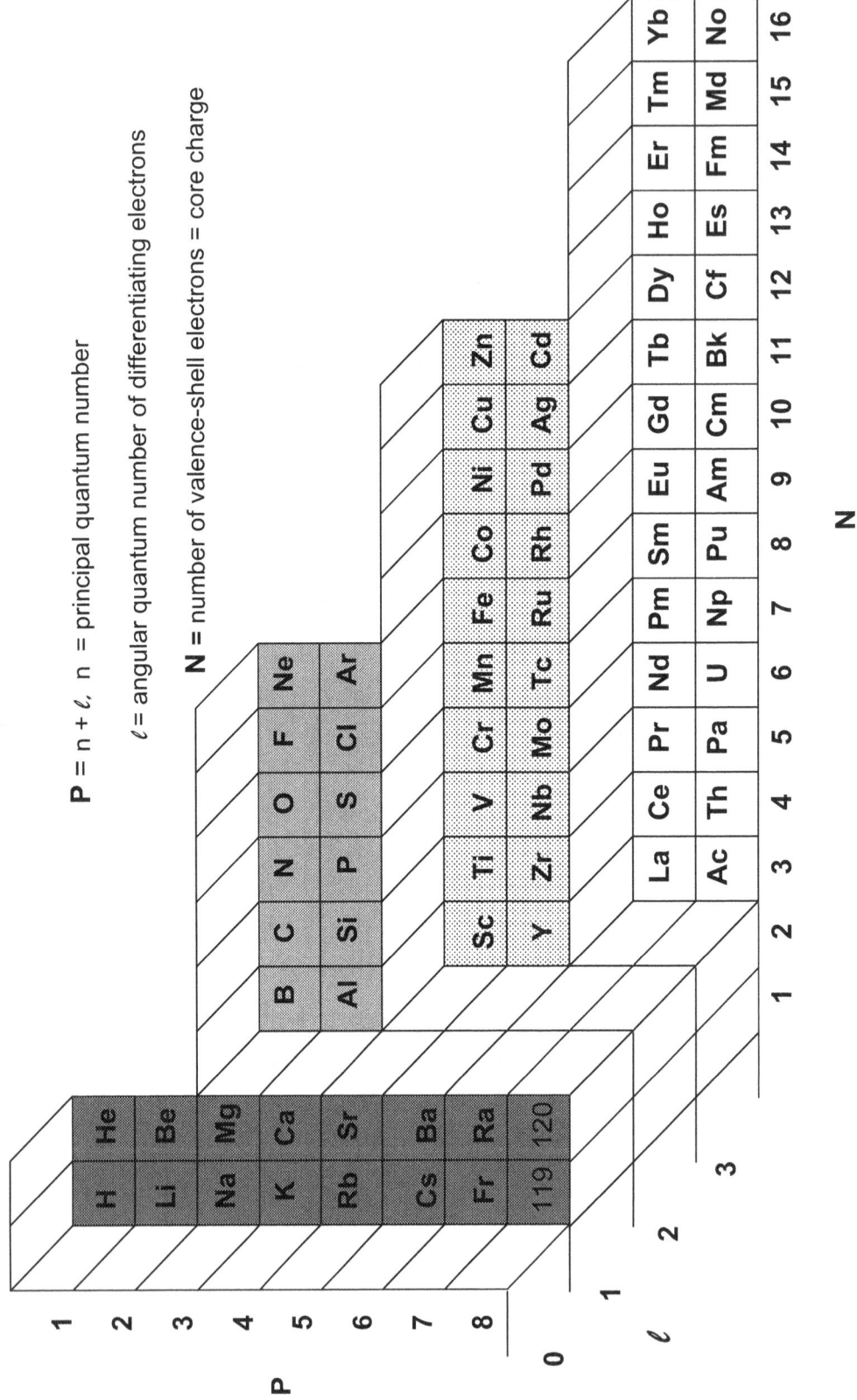

Figure 50. An arrangements of atoms in PℓN space.

The vertical coordinate in Figure 50 is P_{fdps} ($= n + \ell$). The Table highlights graphically (as does the Conventional Periodic Table) the special character of the *s*-block (Appendix VII). Attractive features of the Classification of Atoms by Numerical Values of P, ℓ, and N are listed in Table 14.

Table 14. Attractive features of the P𝓁N arrangement of atoms:
- congeners in vertical, gapless columns
- adjacency for leading secondary kinships (Al-Sc, Si-Ti, P-V, S-Cr, Cl-Mn; Y-La, Zr-Ce).
- members of Newlands-type Groups in vertical P𝓁 planes, N constant
- members of blocks in vertical PN planes, ℓ constant
- proper locations for the problem elements: H, He; La, Ac; and Mg
- prominent locations for the "full-shell" atoms He, Ne, Zn, and Yb.
- prominent display of the distinctive *s*-block
- natural use of natural column labels
- visual display of two instances of the block-length rule:
$$\Delta(\text{block-length})/\Delta\ell = \Delta[2(2\ell+1)]/\Delta\ell = 4$$
- visual display of the block-height rule: $h = 8 - 2\ell$

Not shown clearly in Figure 50 is the order of increasing atomic numbers and, hence, Periodicity's periods. Not shown at all are third and subsequent rows in the *p*- and *d*-blocks. That deficiency can be rectified by stretching Figure 50's ℓ-coordinate, Figure 51 (following page).

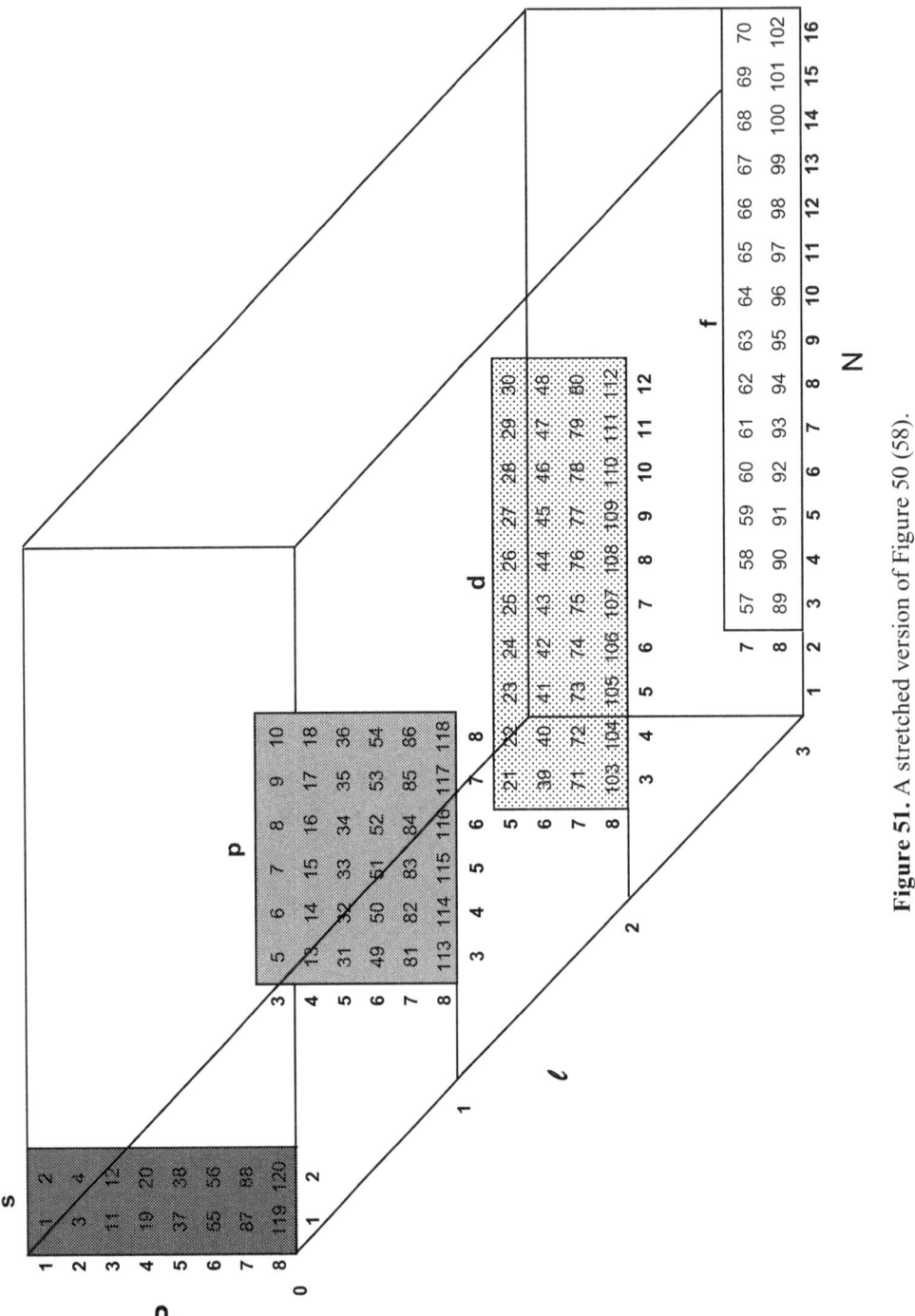

Figure 51. A stretched version of Figure 50 (58).

Order of row occupancy in Figure 51 is given, as usual, by the $n+\ell/\ell$ Rule: smallest $n+\ell\,(=r+2\ell)$ first and, for given $n+\ell$, largest ℓ first. The Rule holds for all periodic tables. Its application is easiest for the LSPT, where values of $n+\ell$ are equal to the ordinal numbers of the Table's periods.

Use of the number of valence-shell electrons e in outer s-, p-, d-, and f-orbitals for a rectangular coordinate, rather than N (the *total* number of valence-shell electrons), slides the p-, d-, and f-blocks two units to the left and aligns them, at the outset, with the s-block. Produced is Figure 52, "A Front-Step Arrangement of Atoms".

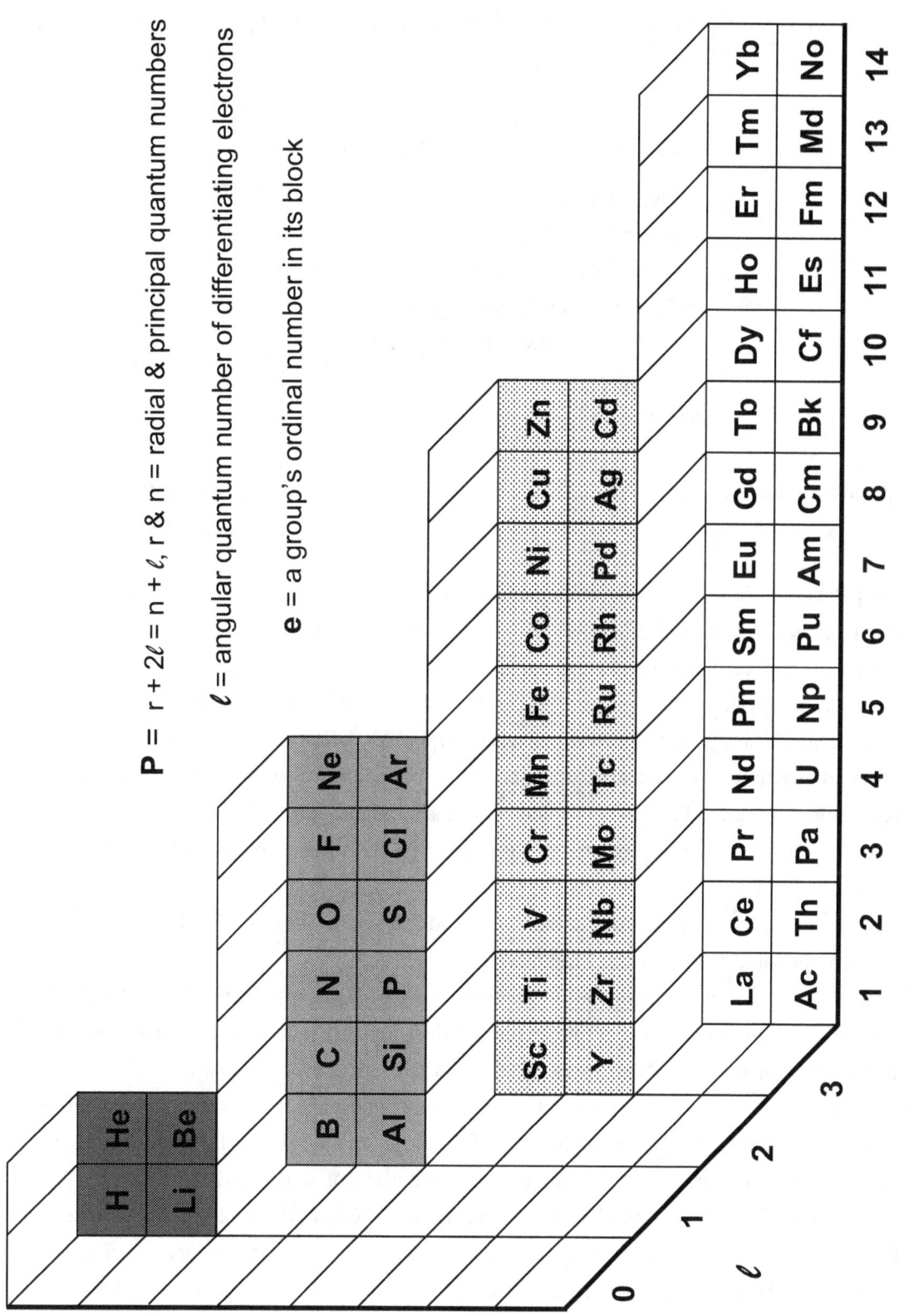

$P = r + 2ℓ = n + ℓ$, r & n = radial & principal quantum numbers

$ℓ$ = angular quantum number of differentiating electrons

e = a group's ordinal number in its block

Figure 52. "The P$ℓ$e" or "Front-Step" Arrangement of Atoms.

The Front Step Arrangement is one of some seven arrangements of atoms of perfect regularity, Table 15.

Table 15. Periodic tables of perfect regularity.

Left-Step Table (Figure 1)
Right-Step Table (Figure 55)
Front-Step Table (Figure 52)
Step-Pyramid, fdps Table (Figure 35)
"Color"-Numeric R/*e* Table (Figure 47)
Row-Centered Table (Figure 48)
"Short-Form", R-Ordinate Table

All periodic tables of perfect regularity have hydrogen above lithium and helium above beryllium.

99. Best Periodic Table? Because analogies among the elements are *many-sided* (Mendeleev), no periodic table is superior to all other periodic tables in all respects. "There is no single best form of the periodic table since the choice depends on the purpose for which the table is used" (1, 3 p136). The question "Which periodic table is best?" is like the question "Which table of data in a Handbook of Chemistry and Physics is best?" "Best for what purpose(s)?" Display of: Chemical valencies? Trends in electronegativity? Atomic structure? Secondary Periodicity? Secondary Kinships? Tertiary Kinships? Gapless periods? Periods' complements of shells and subshells? Periodicity's dyadic character? Madelung's Rule? Locations of "problem elements"? Block-to-block trends? The unique character of the *s*-block?

No periodic table has all the features listed in Appendix XV. The question "Which periodic table is best?" is as impossible as unnecessary to answer.

All real advantages of the many diverse tabulations of the periodic system, remarked Paneth (59), prior to recognition of most of the distinctive features of the LSPT, are incorporated in one or the other of two simple tables: Mendeleev's "short-period" table and a "long-(or medium-long)period form" "We recommend," continued Paneth, "having both simple tables at hand and not to bother with complicated schemes which try to unite the specific advantages of the short- and long-period representations with the help of elaborate geometrical designs, colours, movable parts, or steric models."

The Conventional Periodic Table has the 10 most *conventional* "desirable features" listed in Appendix XV. With helium-above-beryllium, it acquires 5 additional desirable features. The Left-Step Table has 5 of the 10 conventional features and 29 additional features, hence, altogether, 34 of the 45 cited (sometimes interrelated) features. The LSPT's best complement, as regards possession of "desirable features", is a Step-Pyramid Table, *s*-elements on the left, Figure 53 (following page), discussed in Appendix XVI. Jointly the two tables embrace all of the "desirable features" except the last two cited in Appendix XV.

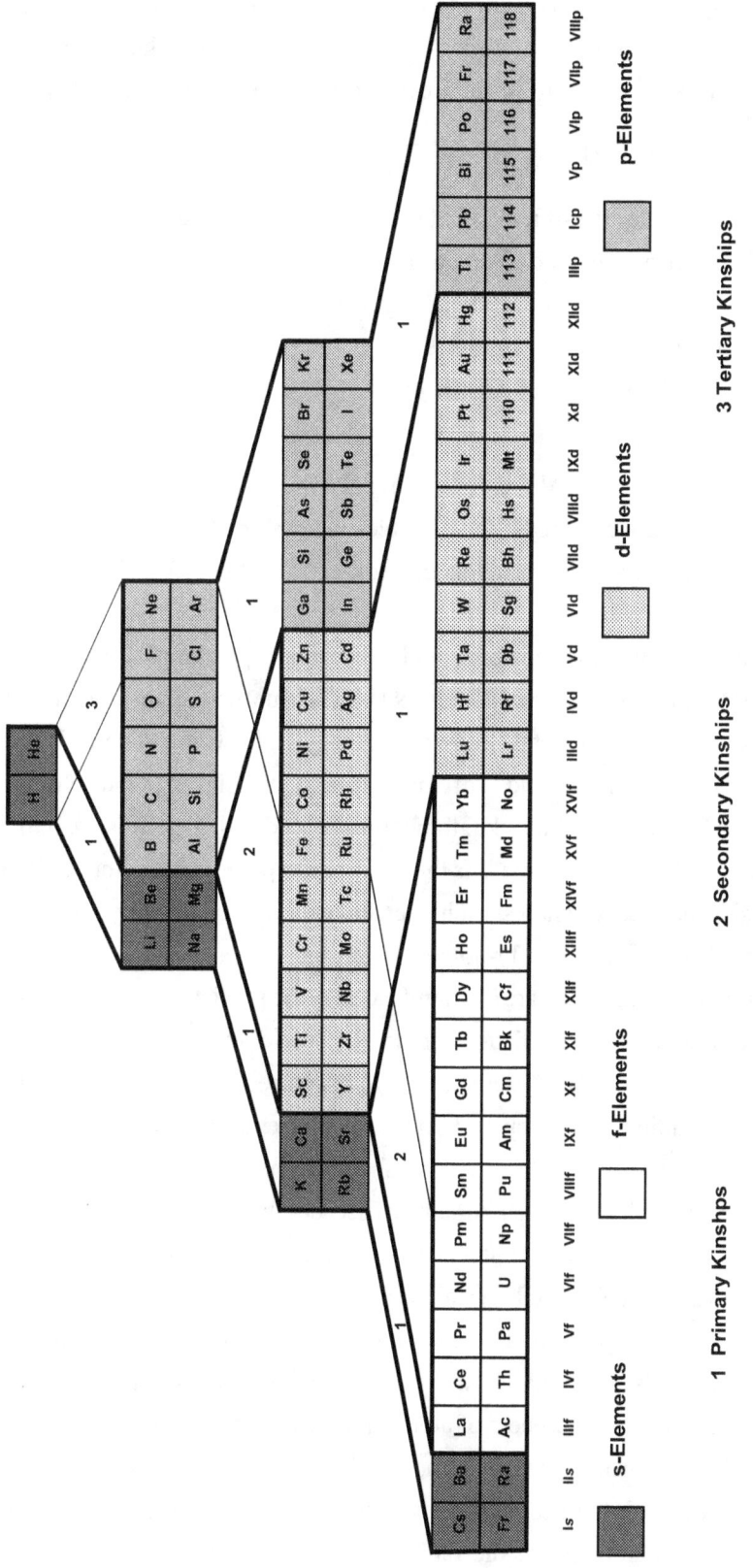

Figure 53. The sfdp Step-Pyramid Periodic Table.

Each of the 26 desirable features cited in Appendix XV that "requires He/Be" may be viewed, alternatively, as an argument for placement of helium above beryllium in periodic tables.

Related to the titular question of this section — and having the same answer — is the question: Which vertical coordinate is the best one for a periodic table? Possible answers are listed in Table 16.

Table 16. Possible vertical coordinates for periodic tables.

1. Number of occupied electron shells
2. Numerical value of the Madelung parameter: $n + \ell$
3. Row-orbital ordinal number, R
4. None of the above (cf. Figure 45)

The first choice in Table 16 yields the Conventional Periodic Table, the second one the Left-Step Table. When their periods are centered, the two tables yield Step-Pyramid Tables, *s*-elements on the left and on the right, respectively. Hybrids of the first and third choices yield Mendeleevian "short form" tables.

Most periodic tables may display any one, or all, of several alphanumeric column labels: European AB, American ABA, and Nℓ and/or *e*ℓ labels (**94**). A virtue of the European AB scheme is its association with a classification of metals known, variously, as the Chatt-Arland-Davies "class a" and "class b" metals, the Pearson "hard" and "soft" metals, and the original, Goldschmidt, geochemical, "lithophilic" and "chalcophilic" metals, aka, to beginning students of chemistry, in the days of "wet" chemistry and "stinking hydrogen sulfide", as the "hydroxide-precipitated" and "sulfide-precipitated" cations of the classical qualitative analysis scheme, and called more recently "p6" and "d10" metals: i.e., metals with "noble gas" and "pseudo-noble gas" cores.

Hallmarks of useful periodic tables are the significance of the predictions they give rise to in their time. Gaps in his tables led Mendeleev to predictions of the existence of undiscovered elements and what their leading properties might be. Misplaced elements led to his corrections of atomic weights, most notably those of beryllium and uranium. Many topics discussed in this report may be recast as implications of the shape of the Left-Step Periodic Table. Because of its shape and because of its general neglect by chemists and physicists, it has been for the present author for two decades the best periodic table for suggesting fresh lines of inquiry regarding the Periodic System. It's suggested, e.g., existence of: block-to-block trends; rules of triads, group sizes, and full shells; a staircase inequality; novel atomanalogies; the phenomenon of secondary periodicity; a category mistake regarding helium's location above neon; a rationalization of the magnesium issue; two sets of natural column labels (Nℓ and *e*ℓ); an isomorphism between Periodicity's integers and the leading quantum numbers of atomic physics; a qualitative physical explanation for the shapes of periodic tables; mathematical and physical insights into Madelung's diagram; and Periodicity's Fibonacci character. Although the time has passed for considering the following implication to be a prediction, the shape of the Left-Step Periodic Table implies that a physical model that accounts for leading features of Chemical Periodicity will probably exhibit a high degree of regularity. Finally, the author is indebted to his brother for pointing out that a new location for helium in periodic tables implies profitable times ahead for publishers of books that display obsolete periodic tables.

100. Topologically Equivalent String Figures. Arguments in this report in support of the assertion that helium belongs above beryllium in *all* periodic tables are incomplete. The assertion is an extrapolation, from a feature of perfectly regular periodic tables to a suggested feature for less regular periodic tables.

In several otherwise perfectly regular periodic tables, helium-above-neon makes helium and/or hydrogen stand out like an "angry digit" (60) — if not in all periodic tables. In, e.g., a Sanderson-like, double-footnoted periodic table (Figure 2, lower right table), the table's shape at its top, with hydrogen above lithium, appears to demand, on grounds of symmetry, placement of helium above neon.

$$H \qquad\qquad\qquad\qquad He$$

$$Li \quad Be \quad B \quad C \quad N \quad O \quad F \quad Ne$$

For the Left-Step Periodic Table, on the other hand, the argument from regularity leads to placement of helium above beryllium. For Sanderson's table the regularity completed by He/Ne is, however, merely *local*, whereas for the LSPT the regularity completed by He/Be is *global*. Accordingly, it's been a leading tenet of this report that placement of an element in the Periodic *System* should take into account the character of the *entire* System, not merely a small part of it.

> The notion that an element (such as helium) can be correctly placed in a tabular expression of the Periodic *System* merely by considering the properties of three simple substances (helium, neon, and beryllium) is naïve in the extreme. It ignores completely over 97 percent of the System, not to mention, for instance, Mendeleev's "absolute distinction"; the phenomenon of First-Element Distinctiveness, *particularly in the s- and p-blocks;* rules of triads, group size, and full shells; the extensive data of atomic spectroscopy, and its interpretation; and a correspondence between ordinal numbers of Periodicity's blocks and blocks' rows and the angular and radial quantum numbers of atomic physics.

Raised, once again, is the issue of the "angry digit" irregularity of conventional periodic tables. The perfect regularity of perfectly regular periodic tables, because it is a characteristic of only a few tables, limits, logically, the generality of conclusions based upon it. That limitation would not be significant if the periodic tables of perfect regularity could display graphically all of Periodicity's important features. But they do not.

> "In physics," points out a physicist, "violations of symmetry are viewed as more beautiful than perfect symmetry (as is the case for 'beauty marks' on a woman's face and asymmetries in art and in nature all around us). Maybe that is why a 'perfect' PT [perfect in regularity and perfect in expression of all known chemical kinships] is not possible — because nature is more beautiful than that! . . . Beauty is not the same as perfect symmetry (or perfect regularity in a PT?). If there were just one perfect PT,

111

chemistry might be pretty boring." There'd be no "helium problem", no "column-labeling controversy", and no "fresh energy" (for the Periodic Law) to report on.

Required for display of all desirable features that periodic tables might display (Appendix XV) are several periodic tables, some of which lack perfect regularity and, hence, may not require, on logical grounds, placement of helium above beryllium. Required for rigor of reasons for relocating helium in periodic tables is a way of looking at periodic tables that suggests that helium-above-neon, if not an angry digit in all periodic tables, is at least, in some sense, an unusual stretch (m).

The word *stretch* brings to mind topology. Periodic tables are, in one sense, like the proverbial doughnut and coffee cup. To the casual eye they look different. To a topologist they are the same thing: in the case of the doughnut and cup, a hole, through which one can stick a finger; in the case of periodic tables, string-figure variations on Mendeleev's Line, Figures 54, through which one might demonstrate a topological equivalence.

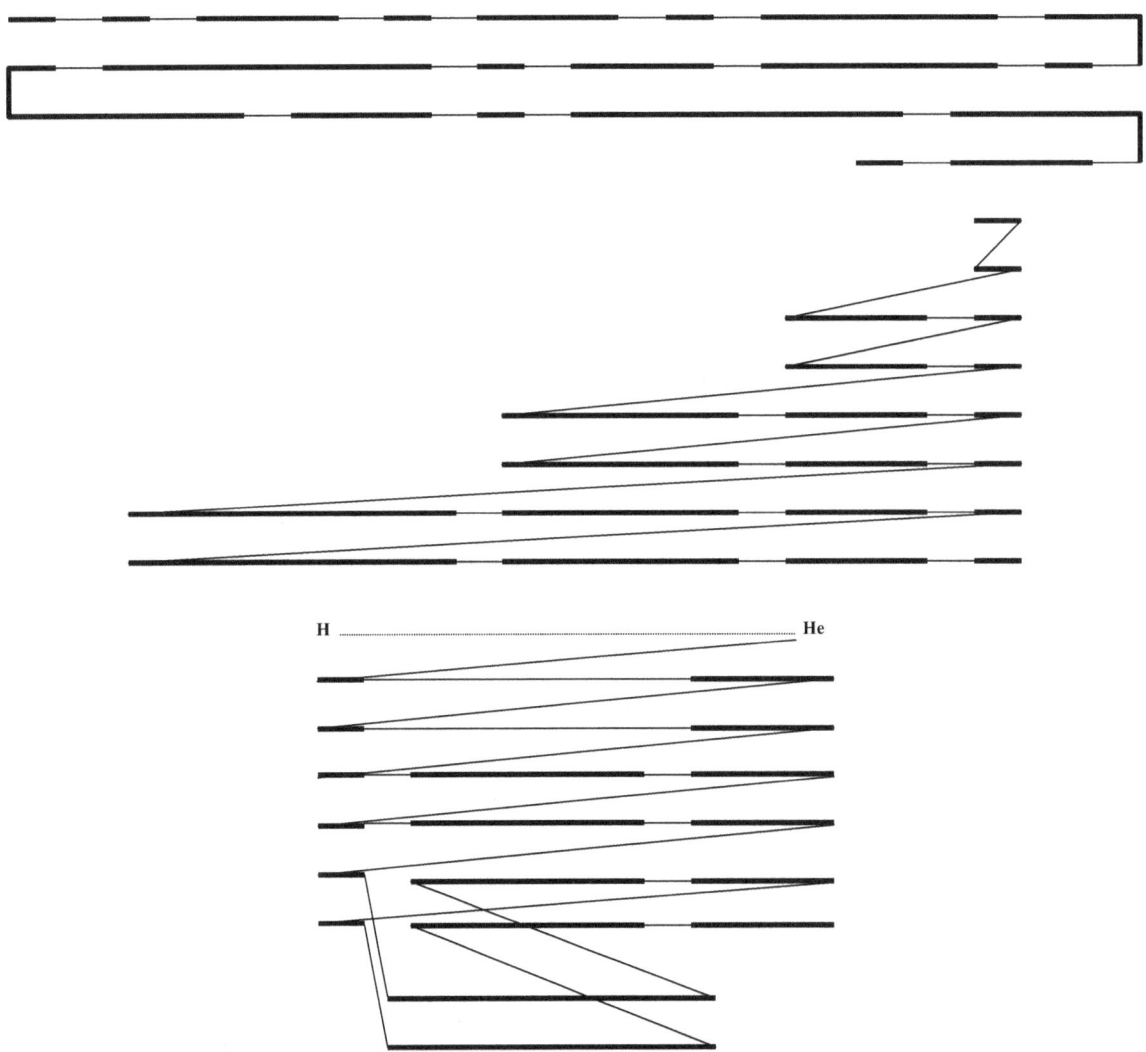

Figure 54. The string-figure character of topologically equivalent periodic tables.

Stretching straight out the second and third string figures in Figure 54 and then shrinking the long (diagonal) tie-lines to the same lengths as the other tie-lines yields one and the same figure: "Mendeleev's Line", in a straight line.

The Bottom Line: Any periodic table can be transformed, topologically, into Mendeleev's Line, which can be transformed, topologically, into any periodic table.

All periodic tables are topologically equivalent to each other.

Figure 55 below is a string-figure representation of the "Right-Step Periodic Table".

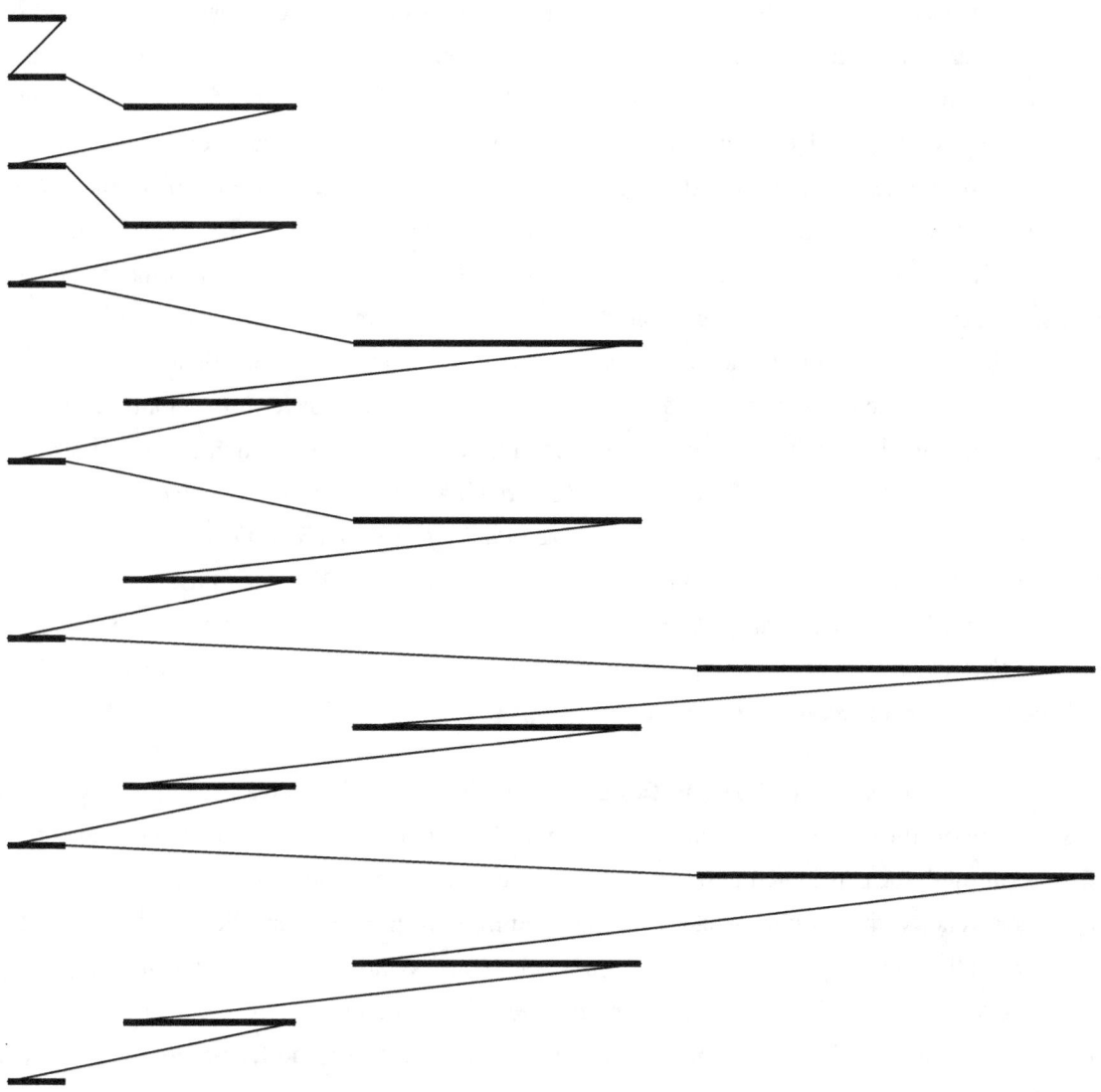

Figure 55. String-figure representation of the Right-Step, sfdp Periodic Table.

Justification on the left of the Right-Step Table's *p*-, *d*-, and *f*-segments yields the string-figure representation of Figure 47. The Right-Step Table is another table of no irregularities. It is an answer to the question: Can blocks' elements be arranged in accordance with their atomic numbers and the

Periodic Law in a perfectly regular manner in the order s *p d f?* Yes, albeit with gaps in columns and periods. Since, however, the gaps occur in a *perfectly regular manner,* they have — unlike the famous gaps in Mendeleev's early tables (but like the traditional, irregular gaps in the Conventional Periodic Table) — no chemical significance.

Topological equivalence of irregularity-free and non-irregularity-free periodic tables brings -

<u>A New Dimension to the Helium-Placement Problem:</u>

The distance in tabular expressions of periodicity between H and He.

In Mendeleev's Line no interruptions occur between elements that have consecutive atomic numbers. H and He are adjacent to each other. In periodic tables interruptions occur where new periods begin. For the Conventional Periodic Table those interruptions occur on occupancy with increasing atomic number of an orbital that has a new (larger) value of n. For the LSPT the interruptions occur on occupancy of an orbital that has a new (larger) value of $n + \ell$. Neither parameter, n nor $n + \ell$, changes, however, on passage from hydrogen to helium. True, maximum oxidation number declines, from +1 to 0. Yet such declines occur at the extended ends of the first rows of *all* blocks. An absence of such a decline at the end of the first row of the *s*-block would make that block in that regard anomalous. Also, reactivity declines dramatically in passage from H to He. But, again, that kind of behavior occurs, also, in the *p*-block, on passage from fluorine to neon. In summary, the dramatic difference in chemistry between hydrogen and helium is not a legitimate reason for distancing them from each other in periodic tables. The Conventional Table's interruptions occur on passage from a nonmetal (a noble gas) to a metal (an alkali metal), not from a nonmetal (hydrogen) to another nonmetal (helium).

In all irregularity-free periodic tables, hydrogen and helium are, as in Mendeleev's parent Line, adjacent to each other. Passage from the Line to irregularity-free periodic tables creates no change in the hydrogen-helium distance. Passage from the Line to the "long-form" of the Conventional Periodic Table, helium above neon, creates, however, a stretch of unprecedented proportions between two elements that are in the same period and that have their valence-shell electrons in the same subshell and that have successive atomic numbers.

101. The "The" of "The Periodic Table". Although topological equivalence of periodic tables does not support the conventional placement of helium above neon, it does support the conventional use of what might be considered to be, in view of the existence of some 700 periodic tables, a naïve, linguistic infelicity: the definite article "the" in the familiar phrase "the periodic table". For in a deep sense (**100**) all periodic tables are the same thing: elements' symbols (on an elastic string) grouped (via string-stretches) by Groups (the same in all tables) grouped, sometimes, by blocks (in, e.g., the Conventional Periodic Table), grouped, in two instances, by size [in the Left-Step and Front-Step (Pℓe) Periodic Tables].

Reasons, and Refutations, Regarding He/Ne

The question Why is helium placed above neon in periodic tables instead of above beryllium? usually generates one or more of the following *responses.* The accompanying refutations transform arguments for He/Ne into arguments, it is suggested, for He/Be.

(1)   *Helium and neon are inert gases.* Implicit in that response is the assumption that vertical adjacency in periodic tables always implies similarity. By the Mendeleev-Jensen Rules of First-Element Distinctiveness, however, the cited similarity between helium and neon is a powerful reason for <u>not</u> placing helium above neon in periodic tables. Moreover, the "inert gas" argument ignores Mendeleev's "absolute distinction". The Periodic Law is about *atoms,* not *simple substances,* such as "inert gases". To repeat:

> Classifying helium with neon is an *artificial* classification of *simple substances.* Classifying helium with beryllium is a *natural* classification of *atoms.*

Both types of classification, noted Mendeleev, are useful. Only the Periodic Classification of Atoms, however, is, in Mendeleev's view, a natural classification. Insertion of an artificial classification into it is a category mistake.

(2)   *Helium is not, like beryllium, a metal.* Neither is carbon a metal; yet it is grouped, nonetheless, by the Periodic Law, with the metals tin and lead. Nor is boron a metal; yet it is grouped, even so, with metallic aluminum, gallium, indium, and thallium. And hydrogen, another nonmetal, is usually grouped, nevertheless, after Mendeleev, with the alkali metals. Uneasily sits helium, however, in a chemist's mind, atop a column of metals, until it's realized that the metal-nonmetal dichotomy may be irrelevant (**19**). Rejection of He-above-Be because helium is not a metal and beryllium is is another category mistake.

(3)   *Helium cannot be oxidized.* Viewed in isolation, above the easily oxidized alkaline earth metals, helium looks, at first glance, out of place. Viewed in the broader context of the <u>entire Periodic System</u>, helium, in its inertness, looks more natural above beryllium, in the sequence of increasing oxidizability of the first fully occupied 1s, 2p, 3d, and 4f subshells of He, Ne, Zn, and Yb, than it does above neon (in some such sequence as Be, He, Zn, and Yb).

(4)   *Helium, like neon and the other noble gases, is monatomic.* Also monatomic in the gas phase are the alkaline earth metals and the volatile metals, zinc, cadmium, and mercury. Monatomicity, alone, does not locate helium in the Periodic System.

(5)   *Helium, like neon and the other noble gases, has a negative electron affinity.* Also having negative electron affinities are the alkaline earth metals and the volatile metals. That property, alone, does not locate helium in the Periodic System.

(6)   *Helium, like neon and the other noble gases, is followed in the Periodic System by an alkali metal.* But, also, hydrogen, like fluorine, is followed, in two steps, by an alkali metal. But it has not,

therefore, a primary kinship with fluorine. In all periodic tables the first period is unique. Only for the first period is an exit from a period followed by immediate reentry into the same block.

(7) *Helium, like neon, has a full outer shell.* Helium and neon are, in fact, the only two atoms in the Periodic System that have full outer shells. Ar, Kr, Xe, and Rn do not. Shells 1, 2, 3, and 4 (K, L, M, and N) are first fully occupied at the ends of the first rows of the first periods of dyads 1, 2, 3, and 4, at He, Ne, Zn, and Yb (Figure 38). Having a full outer shell is, therefore, not a compelling reason for placing helium in the Noble Gas Group.

(8) *The Left-Step Periodic Table, which places helium above beryllium, is a "spectroscopic table" or "electron orbital filling chart", not a chemical table.* As shown in Figure 1 and Table 1, however, the LSPT can be captured from chemical data.

(9) *Helium has no chemistry. Hence, no chemical evidence supports helium's placement above beryllium in periodic table.* Helium's nobility might be regarded, however, as its most distinctive — if only — chemical property. It completes, as noted, a chemical trend in nobility of ℓ-subshells on first full occupancy, starting with the oxidizable 4f-subshell of Yb, continuing with the barely noble 3d-subshell of Zn, and ending with the highly noble and most noble 2p- and 1s-subshells of, respectively, Ne and He.

(10) *Classification of helium with beryllium in the Periodic System is merely a theory, not a fact.* That assertion is, in Pauli's words, "not even wrong". "A fact," observed the polymath William Whewell (61), "is a familiar theory" (n).

From scientific, philosophical, and historical points of view, a compelling *chemical* reason for placement of helium above neon in a periodic table would be that it eliminates a gap in periodic tables. Ironically, however, it creates a permanent gap where previously none existed.

In the big picture, no valid, *scientific* reasons exist for placing helium above neon in periodic tables. In the words of Mendeleev regarding the problem for the periodic law of peroxides (22, p218-9), "not only do [arguments against He/Be] not invalidate [it], they fully support it." He/Ne is a *CATEGORY MISTAKE*. It switches categories midstream, so to speak, from *atoms* to *simple substances*. Simultaneously, it ignores the Periodic System's systematic exceptions to the Similarity-Adjacency Rule and violates or destroys most of Periodicity's higher-order regularities.

Helium and the Noble Gases have had an interesting history. On the one hand, helium has been misnamed, misbelieved (62), misclassified, and misused (in not being used as a leading example of light-element distinctiveness). On the other hand, the Noble Gas Group has illustrated Bohr's principle that the opposite of a profound statement may be a profound statement: that, e.g., for every gain there's a loss; and, reciprocally: for every loss there may be a gain. In the middle of the 20th century the Noble Gas Group lost what was considered to be at the time a cornerstone of chemical theory: Noble Gas nobility. Gained was a new field of chemistry and, among additional benefits, unsuspected examples of the Isoelectronic Principle. Lost, now, at the beginning of the 21st century, if the facts and principles set forth in this report are sound, is the Noble Gas Group's traditional first member. Gained, the report suggests, is a new, more expansive view of Chemical Periodicity.

Summary and Concluding Remarks

The answer given here to the Helium Question To Be or Not to Be? is, on both chemical and physical grounds, a resounding Yes! The most noble of the noble gases is not a Noble Gas. Helium's natural position in Periodic Tables is in the *s*-block above beryllium, where it supports many regularities in the Periodic System, hitherto overlooked, owing to helium's traditional location above neon; to a conventionally footnoted *f*-block; to nearly universal use of periodic tables that have the *s*-elements on the left; and, consequently, to nearly universal neglect of implications for chemical thought of the *fdps* Periodic Table.

Full and unambiguous use of the concept of stationary states, said Bohr, was the key to the development of quantum mechanics. Full and unambiguous use of the distinctive features of the Left-Step Periodic Table has been the key to recognition of Chemical Periodicity's commonly overlooked regularities. Their gradual emergence from irregularities of traditional periodic tables brings to mind Aristotle's synopsis of the Iliad and the Odyssey: Woman abducted. Long war. One guy has a hard time getting home. End of account. The rest is episodes. Similarly here: Chemical capture of the LSPT. An Isomorphism with Atomic Physics. Block-to-Block Trends. End of report. The rest is episodes. Crucial in avoiding category mistakes is Mendeleev's "absolute distinction" concerning elements as *atoms* and elements as simple substances.

Atoms are the subject in this report of numerical indices of first-element distinctiveness; block-to-block trends in first-stage ionization energies; trends in refractivities and *ℓ*-nobility; Rules of Triads and Full Shells; a Row-Orbital Correspondence Principle; Madelung's order-of-orbital-occupancy parameter; physical explanation for late occupancy of high-*ℓ* orbitals; physical description of primary, secondary, and tertiary chemical kinships; and resolution of the zinc-magnesium issue and the helium-neon inconsistency.

The role of atoms in understanding Chemical Periodicity brings to mind one of Feynman's favorite questions: If you could leave but one sentence to posterity to express mankind's leading intellectual achievement what would it be? Feynman's answer: *Atoms* exist.

Every thing, all stars (and presumably their planets), is made of the same stuff, the same atoms. If we wished to broadcast to the universe at large a simple message to indicate to possible extra-terrestrial beings existence of intelligent life on earth, what might it be? Something about atoms? This report suggests the following figure from out of the test tubes and minds of inorganic chemists:

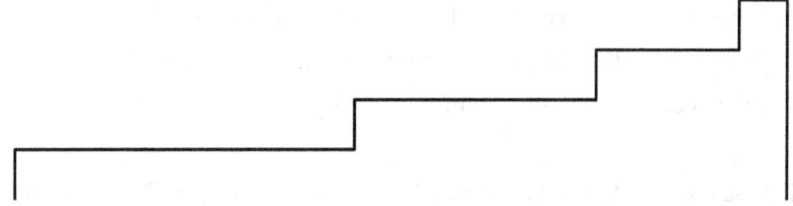

Displayed, figuratively, are the integers 1 3 5 7 and their sums 1 4 9 16. Inorganikers, anywhere, noting that step-height is the same as step-width at the top, might immediately double the integers, to 2 6 10 14 and to 2 2 8 8 18 18 32 32, and conclude: "Aha! Someone else in the universe knows about atoms."

If one could leave but one *paragraph* about Periodic Tables of Atoms (a single sentence, adequate to the subject, would be unwieldy), what might it be? Results reported in this report suggest the statement in Appendix XVII. It's a response, also, to one of J. Arthur Campbell's favorite questions: What's periodic about Periodic Tables?

Classification of the chemical elements *in accordance with the Periodic Law* is a nuanced affair. *Being similar as simple substances* (helium and neon, the rare earth metals, magnesium and zinc) does not mean, necessarily, "in the same group" (although often it does). Similarly, *being different as simple substances* (helium and beryllium; also: hydrogen and lithium, boron and aluminum, carbon and lead, nitrogen and phosphorus, oxygen and sulfur) *does not mean, necessarily* (owing to the phenomenon of First-Element Distinctiveness), *"in different groups"* (although usually it does, since most elements are not "light elements"). Elements may be similar yet not in the same Group (particularly in the case of the rare earths). And elements may be dissimilar yet in the same Group (particularly at the tops of the *s*- and *p*-blocks). Those statements concerning where and in what ways similarities as simple substances do and do not exist among congeners are part of the lore that makes periodic tables useful.

Some lore about periodic tables has hindered development of Periodic System Systematics. PSS has been held hostage, historically, by such linguistic infelicities as the "Group/sub-Group" dichotomy; the phrases "lanthanide and actinide 'series'", "transition metals", "spectroscopic table", "*the* periodic table", "The Periodic Classification of Elements" [considered as simple substances (and sometimes displayed as such in the format of a periodic table)], "the hydrogen-helium 'block'", "the Main Group 'block'"; by expression of the Madelung order of orbital occupancy parameter as "$n + \ell$" (rather than as the physically more perspicuous form $r + 2\ell$); by the European AB column labels (which ignore the *f*-block); by the American ABA column labels (which, like the European scheme, does not distinguish between *s*- and *p*-block elements); by IUPAC's 1-18 scheme (which fails in the all the ways the AB and ABA schemes do and, in addition, fails to indicate secondary chemical kinships); and, last but not least, by the custom of considering helium to be a "congener" of neon.

Helium above neon in periodic tables wreaks havoc with many of Periodicity's regularities. It seems time to consider its retirement. It's served the Periodic System well, during its early years, by calling attention to one of the System's leading features: similarities of (heavy) congeners to each other, as simple substances. At the same time, however, He/Ne diverts attention from regular departures from congener similarity and other important features of Periodicity. Its widespread appearance on walls of chemistry buildings above neon called to mind after near completion of this report the story of the acerbic professor who returned a paper with grade –10, "for an unnecessary display of ignorance."

Ignored by He/Ne are the Mendeleev-Jensen Rules of First-Element Distinctiveness; Mendeleev's Absolute Distinction; Jensen's distinctions regarding types of chemical kinships; the Rules of Dyads, Triads, Group Sizes, and Full Shells; block-to-block trends in elements' properties, particularly first-element distinctiveness; the pattern of secondary periodicity in the *s*-block vis-à-vis its pattern in the *p*-block; and a Correspondence regarding ordinal numbers of blocks' rows in periodic tables and atomic orbitals' radial quantum numbers.

Some skeptics regarding He/Be have considered numerical regularities cited in its support "pure numerology". Mendeleev might have agreed. "As is well known, Mendeleev expected that the secret of the periodic law would probably be discovered through the theory of numbers, and not through the theory of continuous functions, for he saw that 'in chemistry we everywhere see jumps'" (63).

Mendeleev loved regularity. To maintain the checkerboard character of his favorite periodic table up through the table's top series, he placed (Figure 23) boron above scandium, carbon above titanium, nitrogen above vanadium, oxygen above chromium, and fluorine above manganese, all secondary kinships (Figure 35). Had Mendeleev known what we know today, it seems likely to the author of the following conjecture that he would have embraced the perfect regularity of the Left-Step, Right-Step, and Front-Step Periodic Tables, all of which place helium above beryllium. Mendeleev's work on the Periodic Law supports the advice: Seek regularity *and distrust it* — unless it's perfect. Risky are extrapolations of regularities of an irregular scheme that, after the extrapolations, still remains, in some respects, irregular.

Use of more than one form of the periodic table is not an imperfection of chemical thought. It's a step toward its perfection. Understanding, said Wittgenstein, paraphrased [and augmented], is seeing that the same thing said [graphed, or tabulated] different ways is the same thing (64). Nowhere is it engraved in stone: Thou shalt use only one periodic table. The Master himself is said to have considered some sixty periodic tables. In *Principles* Mendeleev displays, twice, two tables, about which he says: "Although this arrangement [an early version of Figure 37, rotated 90 degrees] best expresses the Periodic Law, the distribution of the elements according to groups and series in the table on [the next page: a "short-form" table, Figure 23] is perhaps better" (65).

Best for exhibiting chemically important runs through elements' maximum states of oxidation are conventional, *sfdp*-type tables; for exhibiting secondary chemical kinships, step-pyramid and "short-form" tables; for much else, the Left-Step Periodic Table. The tables complement each other. Each table's limitations highlight distinctive features of other tables. Virtually all tables presently highlight one of the LSPT's most distinctive features: He/Be (66).

What's next for the Periodic Law? "[I]t needs not only new applications," said Mendeleev at the end of his famous Faraday Lecture of 1889, "but also improvements, further development, and plenty of fresh energy" (67). Improvements since 1889 might include relocation of helium in periodic tables. New applications might include uses of the LSPT.

Developments since 1889 include: discovery of the remainder of the rare earths, discovery of the noble gases, Thompson's electron, classification of atomic spectra, Rutherford's nucleus, Moseley's atomic numbers, Soddy's isotopes, Chadwick's neutron, Planck's constant, Schrodinger's wave mechanics, Bohr's Aufbau Process, Janet's Table, Madelung's rule, synthesis of technicium, syntheses of early trans-uranium elements, Seaborg's actinide hypothesis, synthesis of Mendeleevium, additions to the bottom of the *d*-block, accurate calculations of atomic energy levels, and calculations of relativistic effects. Within a century or so computer scientists may be able to calculate accurately properties of atoms with atomic numbers up to what, 1000? Today's periodic tables sit atop a huge manifold of unknown information that, as brought forth, may provide the Periodic Law with "plenty of fresh energy".

"Fresh energy" for the Periodic Law might help to restore inorganic chemistry to its natural position in the chemistry curriculum, from which it's fallen, precipitously — at the same time, ironically, that the field itself experienced a renaissance. Perhaps never before in the history of chemistry has the gap between the achievements of inorganic chemistry and its role in education in — and through (o) — chemistry been greater.

One's reminded of a question Pauling used to ask: Why aren't metals discussed more widely in beginning chemistry courses? Most elements are metals. Of course, teachers teach what's teachable (historically, the three R's). And chemists, Pauling added, haven't (widely accepted) *chemical* models of metals and metallic compounds (Appendix XVIII).

Today Pauling's question might be: Why isn't *inorganic chemistry* discussed more widely in introductory chemistry courses? The shocking disappearance of names of elements from titles of chapters in a number of recent textbooks of general chemistry and the statement by a critic of an early draft of this report that the average general chemistry teacher's use of the periodic table is "extremely superficial" (2) stem, in part, the author believes, for want of simple yet rigorous, in-house, *chemical* account of inorganic chemistry's central doctrine, owing, in large measure, to helium's illogical location in periodic tables; to use, usually, of only one periodic table — or, if more than one, not one that features blocks in the order *fdps;* and to use of Madelung's diagram alongside the Conventional Periodic Table (to which it bears no direct, logical relation) as merely a mnemonic for writing down atoms' ground state electron configurations.

Three strikes and you're out (of the curriculum). Strike 1: You're merely a "spectroscopic table". Strike 2: You're merely an "order of orbital filling chart". Strike 3: You're merely a "mnemonic crutch".

Chemists' success in teaching inorganic chemistry, Orville Chapman might have said, is no better than their expression of it. The key to a scientifically and heuristically satisfying account of Chemical Periodicity lies, in the author's opinion, in placement of helium above beryllium in all periodic tables, and in use, on occasion, of the Left-Step Periodic Table. By and large, however, this report has not been about whether or not one *should* accept He/Be, all the time, and use, at times, more than one periodic table. Those are trans-scientific issues. The report reports what happens *if* one transposes Chemical Periodicity into that new key, the key of "Be", the natural signature of the Left-Step Periodic Table. Produced, in words of Werner and Mendeleev, are new ideas in inorganic chemistry from fresh energy for the periodic law.

It has not escaped the author's notice that the subject of Chemical Periodicity, approached in the manner sketched in Figure 1, is highly teachable. It's not rocket science. With scissors and scotch tape and copies of the top part of Figure 1 (or with appropriately programmed computers), beginners at least as early as the fifth grade can execute the cited Construction Conventions and create for themselves from Mendeleev's Line the Left-Step Periodic Table, easily rearranged to the Conventional Periodic Table, Figure 2.

> *Seen at a glance are central features of a central doctrine*
> *of the central science.*

Promotion of that view of chemistry has been the purpose of periodic tables from the outset. In famous footnote 8 of Chapter 1 of Volume II of *Principles*, Mendeleev wrote: "The periodic law and the periodic system of the elements appeared in the same form as given here in the first addition of this work, begun in 1868 and finished in 1871. In laying out the accumulated information respecting the elements, I had occasion to reflect on their mutual relations." This report attempts to show, following in Mendeleev's footsteps, that locating helium above beryllium in periodic tables and ordering in the mind's eye, if not on lecture-room walls, the periodic system's blocks of Groups in the order *fdps* aids reflection on the elements' mutual relations.

In noted footnote 8a of Chapter 1 of Volume II of *Principles*, Mendeleev wrote (with slight editing): "When the periodic law was first established (1869) there were no reasons for even suspecting the existence of the noble gases. They formed sort of a test of the theoretical side of the periodic law. Ramsay showed in 1900 by their atomic weights that they occupy definite positions between the halogens and the alkali metals. They gave brilliant confirmation to the principles of periodicity."

Paraphrasing Mendeleev, one might add: When the noble gases were first placed in periodic tables, there were no reasons for even suspecting a special assignment for helium. Janet's Left-Step Table of 1927 with helium above beryllium formed, therefore, sort of a test of the theoretical side of the periodic law. Physicists showed by spectrum analysis that, indeed, helium occupies a definite position at the head of the alkaline earth atoms. In that position in periodic tables it augments brilliantly the principles of Chemical Periodicity.

Acknowledgments. The author is indebted to Dr. Bruce Conard and professors Lee Allen, Doyle Britton, John Corbett, Roald Hoffmann, William Jolly, Joel Liebman, Bassam Shakhashiri, Emil Slowinski, and Richard Zare for encouragement; to professor R. Bruce King for, in addition, copies of reprints of English translations of Shchukarev's paper on chemical periodicity (55); to professor Frank Weinhold for, additionally, recent improvements and extensions of the author's discussion of Periodicity; to professor Peter Nelsen, for his interest in the periodic table column labeling issue; to professor Derek Davenport for an invitation to participate in a symposium that led to the first draft of this report; to the late professor George Pimentel, for an appointment to the American Chemical Society's Presidential Ad Hoc Committee on Column Labels, which initiated the author's studies of Chemical Periodicity; to Oliver Sachs for friendly correspondence and his chapter on Mendeleev in "Uncle Tungsten"; and to anonymous reviewers for the *Journal of Chemical Education,* whose vigorous resistance over a number of years to the contents of early drafts of various portions of this report have played a role in its evolution to its present form; and to Authorhouse's Senior Designer Christian Kelly, for his professional contribution to "New Ideas". The author is especially indebted to his daughter, Libby, for perceptive editorial insights; to his brother, Robert Bent, for constant encouragement, valuable suggestions, probing questions, and critical aid in completion of Faraday's advice to investigators ("Work. Finish. *Publish.*"); to professor William Jensen, for sharing over a period of two decades many valuable scientific and historical insights regarding Chemical Periodicity (p); and, above all, to his wife, Anne McKnight Bent, for making feasible, if not always factual, for nearly half a century, a Newtonian steady intention of one's mind on a single subject.

Dedicated to memories of Brian Edward Bent, 1960 – 1996, who arranged for his father's first public lecture on Chemical Periodicity, at Columbia University, in 1988, and to Brian's grandfather, Henry Edward Bent, 1900 – 1987, who introduced his son to Dobereiner's Triads and the Periodic Law in his course on freshman chemistry at the University of Missouri, in 1943.

Pittsburgh, PA
April 2006

Aphorisms and Other Summarizing Statements

With a tip of Henry's hat to his daughter Libby Weberg

Helium and Periodic Tables

- Classification of helium with the Noble Gases is a category mistake.

- The noblest of the noble gases is not a Noble Gas.

- The phenomenon of allotropy implies that the Periodic Classification of the Elements is not a classification of the elements as simple substances, such as diamond or graphite, oxygen or ozone.

- The Periodic Law was based in the first instance on two *atomic* properties: *atomic* weights and *atomic* combining capacities, determined by maximum states of oxidation.

- The classification of the elements according to the Periodic Law is a classification of *atoms*.

- The answer to the question: Where does helium belong in periodic tables? is the question: *What atoms are helium's atoms like?*

- Helium-above-neon in periodic tables is an *artificial* classification of *simple substances*.

- Helium-above-beryllium in periodic tables is a *natural* classification of *atoms*.

Classification by Fundamental Properties

- The most useful classification of the elements, in the long run, is by properties that are the most explanatory.

- Properties of atoms are more fundamental than properties of simple substances, which are explained in terms of the former, not the former in terms of the latter.

- Chemists say "Diamond is composed of carbon atoms", not "Carbon atoms are composed of diamond."

Phenomenology's Advantages and Disadvantages

- Phenomenology at its purest says nothing more, nor less, than the truth. It's never wrong.

- For every gain in explanatory power, starting at zero with pure phenomenology, there's a loss in certainty.

- The definition e = Group's ordinal position in its block admits of no exceptions.

- The definition e = number of electrons in orbitals occupied by a block's predominant type of differentiating electrons has numerous exceptions in the d- and f-blocks.

Various Captures of the LSPT

- Direct passage from a graphic expression of Mendeleev's statement of the Periodic Law to a Periodic Table of no irregularities is possible, by means of four table construction conventions.

- The Left-Step Periodic Table is only a step away from the Conventional Periodic Table.

- The LSPT can be captured from oxidation numbers and the Table Construction Convention of Maximum Regularity without knowledge of Group membership.

- The LSPT can be captured from ionization energies and the Maximum Regularity Convention without knowledge of atomic orbitals.

- A classification is deemed *natural* if the arrangement obtained one way is the same as the arrangement obtained another way.

The LSPT's Shape and the "Problem Elements"

- The shape of the Conventional Periodic Table is to some extent an historical accident.

- The shape of the Left-Step Periodic Table is a powerful tool for assigning the "problem elements" to their correct positions in the Periodic System.

- In predicting existence of gap-filling elements and what their properties might be, Mendeleev relied on a Principle of Regularity on a *local scale.*

- In locating "problem elements" in the Periodic System with the aid of the Left-Step Table, one relies on a Principle of Regularity on a *global scale.*

Helium and the Periodic System

- Helium plays a distinctive role in the Periodic System.

- Helium caps off, or anchors, most of Periodicity's block-to-block trends.

- The change He/Ne to H/Be brings the Ne and Be columns of periodic tables into concordance with a number of generalizations regarding the Periodic System.

- According to Heisenberg, a successful revolution in science changes as little as possible.

- Changing the "N" in He/Ne to "B" revolutionizes one's outlook on the Periodic System.

- Helium's line-lean visible spectrum is similar to the spectra of two other s^2 systems, hydrogen and mercury, and not at all like the line-rich spectrum of p^6 neon.

Hydrogen-Helium Consistency

- Logic requires that hydrogen and helium be treated in an analogous manner in periodic tables.

- Hydrogen-above-lithium calls for helium-above-beryllium.

- Not hydrogen-above-fluorine implies not helium-above-neon.

- In their ionization energies and electronegativities, helium stands to beryllium as — to a good approximation — hydrogen stands to lithium.

Triads

- Periodicity's dyadic character implies existence of triads.

- Groups' triads begin with their 2nd, 4th, and 6th elements.

- Groups' lightest elements are not members of triads.

- Helium forms a triad with neon and argon. By the Triad Rules, therefore, it is not in their Group.

- Beryllium forms a triad with magnesium and calcium. By the Triad Rules, therefore, it is not its Group's first element.

- Helium does not form a triad with beryllium and magnesium. By the Triad Rules, therefore, it could be their Group's first element.

- The Triad Rules solve the problem of all the "problem elements": H and He, La and Ac, and Zn.

- In being a consequence of Periodicity's dyadic character, the Triad Rules reflect fundamental, energy-raising — and, hence, orbital-occupancy-delaying — effects of inner-shell screening on electrons in relatively nonpenetrating orbitals of high angular momentum.

Numerology

- To generate atomic numbers of Periodicity's key Group II, add to each other the numbers created by doubling the squares of the sequence 1 1 2 2 3 3 4 4.

- To obtain the atomic numbers of blocks' first elements, add successively the squares of twice the integers 1 2 3 4, starting with $Z = 1$.

- To obtain the atomic numbers of the elements immediately beneath blocks' first elements add to the latters' atomic numbers twice the squares of the integers 1 2 3 4.

- The statement that the highest ordinal number of a block through which a dyad passes is one less than the dyad's ordinal number describes a left-step periodic table.

- A period's length in the LSPT is twice the square of the ordinal number of its dyad.

- A Group's size, for $Z_{max} = 120$ (and helium above beryllium), is eight less twice the ordinal number of the Group's block.

- The ordinal number of an element at the end of dyad D is -

$$(2/3)D(D + 1)(2D + 1)$$

- $Z_{average}$(block ℓ's first row) $= 4\,\ell\,(\ell^2 + 2)/3 + 2\ell^2 + 1.5$. For the $\ell = 0$ s-block, $Z_{average} = 1.5$.

- Ordinal numbers R of blocks' rows in the Periodic System, in terms of their coordinates P and ℓ in the LSPT, are give by the expression –

$$R = (P^2 + 2P + \delta)/4 - \ell$$

where $\delta = 0$ or 1 for P even or odd.

- Ordinal numbers G of Groups in the LSPT, starting at 1 at the right, are given by the expression –

$$G = 2\ell^2 + 4\ell + 3 - e$$

- The number of spin-orbitals associated with shell n is $2n^2$.

- Dyad D and shell n = D run through the same values of ℓ.

- In explaining Periodicity's "magic" numbers (2, 8, 18, . . .), given by the expression $2D^2$, by reference (instead) to the expression $2n^2$, Pauli got the right answer (the Pauli Exclusion Principle) for the wrong reason.

Different Types of Chemical Kinships

- Periodic tables' primary purpose is exhibition of primary kinships, secondarily secondary kinships.

- Primary kinships are usually indicated by verticality in periodic tables.

- Secondary kinships are indicated by tie-lines in step pyramid tables, adjacency in "short-form" tables, and alphanumeric column labels in block-type periodic tables.

- Step-pyramid and "short form" periodic tables and alphanumeric column labels exist for the same reason: existence of secondary chemical kinships.

- Atoms are related by secondary kinships if they are in different blocks, the same distances from their blocks' *left* ends, and have the same *maximum* states of oxidation.

- Atoms are related by secondary kinships only in their maximum states of oxidation.

- Atoms are related by tertiary kinships if they are in different blocks, the same distances from their blocks' *right* ends, and have the same *minimum* states of oxidation.

- Leading examples of tertiary kinships are H & F, He & Ne.

- Primary, secondary, and tertiary kinships yield primary, secondary, and tertiary triads.

- Most mistakes in periodic tables have been secondary and tertiary kinships mistaken for primary kinships.

- The change from He-above-Ne to He-above-Be doesn't deny a relationship between He and Ne. It redefines it.

- Helium's primary kinship with beryllium is the weakest primary kinship in chemistry. Its tertiary kinship with neon is chemistry's strongest tertiary kinship.

- Chemical kinships may be close, atomically speaking (e.g. He and Be), and yet weak, as regards similarities of the corresponding simple substances. Conversely -

- Chemical kinships may be distant, atomically speaking (e.g. He and Ne), yet strong as regards similarities of the corresponding simple substances.

- The strong, if atomically distant, helium-neon tertiary kinship helps to domesticate the weak, but atomically close, helium-beryllium primary kinship.

The Metal/Nonmetal Dichotomy

- Classification of elements as metals and nonmetals is an artificial classification.

- The requirement that all metals be grouped together in periodic tables is an ad hoc requirement. It is not part of the Periodic Law.

- The trend in the Conventional Periodic Table from basic- to acidic-oxide-formers is irregular across the d-block.

Linguistic Infelicities

- The LSPT leads to recognition of several misleading expressions regarding the Periodic System.

- The "transition elements" are not transition elements in all periodic tables.

- With helium above beryllium, one may say that at the right end of the first row of each block stands an element whose maximum oxidation number is less than its row's maximum oxidation number. It is a "transition" element, in the sense that Mendeleev used that word.

- The classical Group/Subgroup dichotomy is obsolete.

- The Periodic System contains, for Z through 120, *four* types of Groups, by size.

- The phrases "long form", "medium long form", and "short form" betray a misunderstanding of the Periodic System.

- All modern periodic tables contain (at this time, 2006) 32 Groups.

Concealments

- The Conventional Periodic Table conceals most of Periodicity's regularities, owing to three irregularities: location of helium above neon; one block footnoted; and the other blocks arranged, by size, in an irregular manner.

- The Left-Step Periodic Table, unannotated, conceals Periodicity's dramatic changes in elements' properties on entering the *s*-block.

- No periodic table exhibits all of periodicity's important features.

- Which periodic table is best? is like asking Which table utensil is best?

- Chemistry is enriched by having more than one periodic table and more than one set of useful column labels.

- The international controversy regarding IUPAC's 1-18 column-labels was a tempest in a teapot.

Elements' Distinctiveness

- All elements are individuals, some more so than others.

- Relative to their congeners, elements of small atomic weight have highly distinctive properties.

- First occurrence with increasing atomic number of each particular type of feature of periodic tables is particularly distinctive.

- Relative to their congeners, the most distinctive elements are the two lightest elements.

- When located in the *s*-block above beryllium, helium is, as congeners go, the Periodic System's most distinctive element.

- Within Groups, congener distinctiveness increases upward.

- Within the LSPT, distinctiveness of block's first-row elements increases from left-to-right.

- Within blocks distinctiveness of first-row elements tends to increase from left-to-right.

- Broadly speaking, congener distinctiveness increases in the LSPT from lower left to upper right.

- The pattern of congener distinctiveness in the format of the LSPT mirrors, qualitatively, the pattern of energy levels of the hydrogen atom.

- Fractional changes in radii of cores of adjacent congeners serve as Numerical Indices of Elements' Distinctiveness and quantify chemical intuition.

- By Mendeleev's Rule of Light-Element Distinctiveness, two light elements alike as simple substances cannot belong to the same Group.

- The Mendeleev-Jensen Block-to-Block Trend in Light-Element Distinctiveness, although almost universally overlooked, is, nonetheless, arguably, the most general feature of the chemistry of the elements after the Periodic Law itself.

- The Mendeleev-Jensen Rule, together with the electronic interpretation of periodic tables, makes helium-above-neon one of the leading contradictions in the history of chemical thought.

Periodicity's Integers

- Latent in Mendeleev's Line are several sets of integers.

- Leading integers generated by Chemical Periodicity have the same mathematical properties as leading quantum numbers of atomic physics.

- The natural numbers are to mathematics as the elements' ordinal numbers in Mendeleev's Line are to chemistry.

- Just as all of mathematics can in principle be based by logic on the natural numbers, so, too, all of chemistry can in principle be calculated by quantum mechanics from the elements' ordinal numbers in Mendeleev's Line.

Physical Significance of Periodicity's Integers ℓ, r, and n

- According to a Block-Orbital Correspondence, blocks' ordinal numbers ℓ in the LSPT are equal to the angular quantum numbers of the blocks' predominate type of differentiating electrons.

- According to a Row-Orbital Correspondence, rows' ordinal numbers r within their blocks in periodic tables are equal to the radial quantum numbers of the rows' predominant type of differentiating electrons.

- According to the Block- and Row-Orbital Correspondences, Periodicity's Index of Discontinuity n $(= r + \ell)$ is equal to atomic physics' principal quantum number.

- Helium-above-neon in periodic tables creates two columns that do not obey the Rules of Triads, Group Size, and the Row-Orbital Correspondence Principle.

- ℓ-orbitals have a nodal point at the origin of power ℓ.

- An orbital's angular quantum number ℓ is equal to its number of nucleus-containing nodal surfaces.

- An orbital's radial quantum number r is equal to its number of spherical nodal surfaces, counting the one at infinity.

- An orbitals' principle quantum number is the sum of its radial and angular nodal surfaces.

- An orbitals' principle quantum number n is, also, the sum of its degenerate ($r = 0$) and nondegenerate ($r > 0$) spherical nodal surfaces.

Orbital Energies and the Parameter $n + k\ell$

- Orbital energies of one-electron atoms increase with n but not with increases in r or ℓ, n constant.

- For few-electron atoms, orbital energies increase with ℓ, n constant (and, hence, with a corresponding decrease in r), but not as rapidly as with n.

- For many-electron atoms, an increase in orbital energy with ℓ, n constant, may be greater than its increase with n.

- The orbital occupancy rule smallest $r + (\Delta h)\ell$ first and for given $r + (\Delta h)\ell$ largest ℓ first yields left-step periodic tables of step height Δh.

- The expression $n + 0.67\ell$ yields the observed order of orbital occupancy in Bohr's Aufbau Process through Z = 120.

- The coefficient 2 in the Madelung expression $r + 2\ell$ ($= n + \ell$) is larger than necessary.

- The argument that physics has not explained the periodic table because it has not produced a proof of the Madelung Orbital Occupancy Rule naively faults physics for not doing the impossible.

- The order of excited states of one- and two-valence electron atoms have a Madelung-parameter-type character.

- The expression $n + k\ell$ yields surprisingly well the order of excited states of one- and two-valence electron atoms.

- The empirical value of k in the orbital-order parameter $n + k\ell$ increases with increasing numbers of electrons in an atom and decreases with the atom's net charge.

Periodic Tables' Mathematically Curious Coordinate
and Arrangements of Atoms in Three Dimensions

- Periodic tables' horizontal, alphanumerically labeled coordinates are mathematical curiosities.

- Numbers (or numerals) in periodic table' alphanumerically labeled coordinate *repeat themselves* and have *attached suffixes*.

- Periodic tables' horizontal coordinates are chemical projections onto one dimension of two parameters: N, or *e;* and ℓ.

- Disentanglement of periodic tables' composite Nℓ or eℓ coordinate using cylindrical polar coordinates yields a "Periodic Round Table".

- Disentanglement of periodic tables' composite Nℓ coordinate using rectangular coordinates yields a three-dimensional arrangement of atoms that exhibits the *s*-block in a distinctive position and, by adjacency, secondary chemical kinships.

- Disentanglement of periodic tables' composite eℓ coordinate using rectangular coordinates yields a "Front-Step Periodic Table" of perfect regularity.

Additional Summarizing Remarks

- The only grouping according to the Periodic Law of the chemical elements into Groups that is free of ambiguities is based on a set of Natural Table Construction Conventions, which produce a periodic table of no irregularities.

- The LSPT is the only periodic table that combines verticality of Groups' members with gapless periods and columns.

- The Madelung diagram of the order of orbital occupancy in Bohr's Aufbau Process is the Left-Step Periodic Table in disguise.

- Periodicity's order-of-orbital occupancy has the character of a Fibonacci series (in which each term, after the first two, is the sum of the previous two terms).

- The Madelung diagram's ordinate *n* and the LSPT's ordinate $n + \ell$ are the third and fourth terms of a Fibonacci series whose first two terms are *r* and ℓ.

- Step-pyramid and "short form" periodic tables and alphanumeric column labels exist for the same reason: occurrence of secondary chemical kinships.

- The conventionally footnoted lanthanide and actinide "series" raise troublesome questions. Why are they footnoted? For printers' convenience? For scientific reasons? And which of the fifteen elements pairs of elements, La and Ac through Lu and No, are congener of Sc and Y?

- In the LSPT format, the lanthanum and actinium paired series almost literally point through the *d*- and *p*-blocks to the *s*-block and the need for locating at its top Nature's two lightest elements.

- Occurrence of nonmonotonic rates of change of congeners' properties with increasing atomic number is called secondary periodicity.

- Congruence of secondary periodicity's kinks in plots of atomic properties for *s*- and *p*-block elements along the same abscissa requires use of the ordinal numbers of the LSPT's periods.

- Systems of perfect regularity live dangerously.

- The zinc-magnesium issue provides a critical test of the LSPT.

- Zinc's noble 3d subshell, distinctive in its row in the *d*-block, is not distinctive within the Periodic System as a whole.

- The *s*-block plays a distinctive role in the Periodic System because outer s-electrons are valence electrons in all blocks.

- Location of the *s*-block on the left-hand-side of the Conventional Periodic Table blocks recognition of many of Periodicity's regularities.

- Location of helium above neon in periodic tables negates many of Periodicity's regularities.

Foxes know many things, it's said, hedgehogs one big thing. This report has fox- *and* hedgehog-like character. Many things about the properties of the elements, it points out, point to one big thing: Helium's *natural* location in periodic tables is above beryllium, not neon.

Appendix I

Indignation Regarding He/Be in the Words of Henry Armstrong

Armstrong on the Bragg's Model of NaCl (68). Professor W. L. Bragg asserts that "In sodium chloride there appear to be no molecules represented by NaCl. The equality in the number of sodium and chlorine atoms is arrived at by a chess-board pattern of these atoms: it is a result of geometry and not of a pairing off of the atoms." This statement is more than "repugnant to common sense." It is absurd to the nth degree, not chemical cricket. Chemistry is neither chess nor geometry, whatever X-ray physics may be. Such unjustified aspersion of the molecular character of our most necessary condiment must not be allowed any longer to pass unchallenged … It were time that chemists took charge of chemistry once more and protected neophytes against the worship of false gods; at least taught them to ask for something more than chess-board evidence.

Armstrong Paraphrased. Professor H. A. Bent asserts that "The 'He' in periodic tables is not the helium of common sense impressions … Its position above beryllium is arrived at [in part] by a paired-periods pattern of symbols; it is a result of Mendeleev's 'absolute distinction' between 'atoms' and 'simple bodies', not of a simple association of two similar substances". That statement is more than "repugnant to common sense (impressions)". It is absurd to the nth degree, not chemical cricket. Every chemist knows that helium, like neon, is an "inert gas", not, like beryllium, an active metal. Chemistry is neither geometry, circular ad hoc arguments based on questionable hypotheses, atomic physics, nor idealistic distinctions, whatever chemical physics and philosophy may be. Such unjustified aspersions upon helium's traditional position in our most necessary table must not be allowed to pass unchallenged … Chemists should stay in charge of chemistry and protect students against the worship of false gods; at least teach them to ask for something more than paired-periods evidence.

Appendix II

Pauli's Induction of the Exclusion Principle from an Erroneous Interpretation of Periodicity's "Magic Numbers"

The challenge of interpreting lengths of periodic tables' periods in terms of a physical model led Pauli to his most widely known contribution to physical thought. Writes Pais (69):

> "In his discussion of the anomalous Zeeman effect, [Pauli] had introduced the new hypothesis that the valence electron in alkali atoms exhibits a [classically nondescribable] two-valuedness [called, today, 'spin']. Now he assumed the same to be true in all atoms, so that each electron is characterized by four numbers, n, k[$= \ell + 1$], m, and a fourth one capable of two values. It follows that the number of states for given n is not N [$= n^2$], but 2N, hence 2 for n = 1, 8 for n = 2, 18 for n = 3: the mystical numbers 2, 8, 18 ruling periodic tables emerge!!"

By coincidence!! — since $2n^2 = 2D^2$ (**74**).

To account for the fact that a spatial orbital is occupied by at most two electrons, Pauli famously decreed: "In the atom there can never be two or more equivalent electrons for which . . . the values of all [four] quantum numbers coincide."

Pauli got the right answer, the Pauli Exclusion Principle, for the wrong reason. (Pais appears to acknowledge as much in a footnote.) Pauli doubles (owing to the electron's "classically nondescribable two-valuedness", called 'spin') the number of spatial orbitals associated with a principal quantum number n ($= 1, 2, 3, 4$). The latter number is n^2 ($= 1, 4, 9, 16$). Then he assumed (the Exclusion Principle) that each "spin orbital" (2 per spatial orbital) can be occupied by at most one electron.

Pauli's square numbers (1 4 9 16), to retrace his steps, arise from the well-known mathematical fact that such numbers are the sums of the odd numbers beginning with unity (1 3 5 7), where the odd numbers, in turn, are given by an expression of the form $2\ell+1$ ($\ell = 0, 1, 2, 3$), *which is exactly the form of the expression for the number of orbitals that have an angular quantum number ℓ, known to range for a principle quantum number n from 0 to n-1*; hence the sums, for n = 1, 2, 3, and 4, of 1, 1+3, 1+3+5, and 1+3+5+7. Doubled, they are periodicity's "magic numbers". Q.E.D.? Not by a long shot! Orbitals that determine lengths of periods, greater than 2, don't all have the same principal quantum number (**74**). Orbitals occupied, e.g., along the first period of length 18 in the LSPT are, in order, 3d, 4p, and 5s, n = 3, 4, and 5. Yet they do run through the same ℓ values as do all the orbitals for n = 3, namely 3s, 3p, 3d, for which $2n^2 = 18$ [$= 2(1+3+5)$].

In summary, Pauli's explanation of the Periodicity's "magic numbers" was, in a mathematical sense, to use one of Pauli's favorite phrases, "not even wrong". In a physical sense, however, it seems correct to say that -

> *One of the leading principles of Modern Physics was induced, in the first instance, from a faulty interpretation of the "magic numbers" of Chemical Periodicity.*

Theoretical physics has yet to account for the empirical fact that $2n^2 = 2D^2$ (**76**).

Appendix III

Estimates of Helium's Ionization Energy from "Atomanalogies" Suggested by the Shape of the Left-Step Periodic Table

Let P_i be the numerical value of a physical property of a Group's element i. Then for a Group's first three elements, 1, 2, and 3, the ratio of to $P_1 + P_3$ to P_2 is about the same for all Groups of a block. Application of that empirical observation to the two Groups of the *s*-block and to the last two Groups of the p-block yields the following estimates for the ionization energies of He (**24.587**) and Ne (**21.564**), in eV:

$$(H + Na)/Li \approx (He + Mg)/Be$$

$$\rightarrow$$

$$He \approx (H + Na)(Be/Li) - Mg = \mathbf{24.75}$$

$$(F + Br)/Cl \approx (Ne + Kr)/Ar$$

$$\rightarrow$$

$$Ne \approx (F + Br)(Ar/Cl) - Kr = \textbf{21.53}$$

The shape of the Left-Step Periodic Table leads, also, to the empirical observation that the difference between the first-stage ionization energy of the last first-row element of the s-block, He, and that of the last first-row element of the p-block, Ne, in which valence-shell orbitals are "outer orbitals", more so for He than for Ne, is approximately equal to the difference between the ionization energy of the last first-row element of the d-block, Zn, and that of the last first-row element of the f-block, Yb, in which the blocks' preponderant differentiating electrons reside in "inner orbitals", more so for Yb than Zn: i.e. -

$$He - Ne = 3.0 \text{ eV} \approx Zn - Yb = 3.1 \text{ eV.}$$

In other words-

$$He \approx Ne + (Zn - Yb) = 21.6 + 3.1 = \textbf{24.7}$$

The previous expression amounts to expressing the "is" of the expression "A is to B as C is to D" as a minus sign (A - B = C - D) rather than as, previously, a division sign (A/B = C/D). Both schemes work in symbolic logic. Both schemes are used in the following ratio of differences of differences of ionization energies of the first three members of the first and last Groups in the "outer orbital" s- and p-blocks:

$$\frac{(H - Li) - (Li - Na)}{(He - Be) - (Be - Mg)} \approx \frac{(B - Al) - (Al - Ga)}{(Ne - Ar) - (Ar - Kr)}$$

$$\rightarrow$$

$$He \approx \textbf{24.47}$$

The shape of the LSPT and the shell model of atomic structure suggest another relation among first-stage ionization energies. The ionization energy of helium, a $1s^2$ system, is equal to that of hydrogen, the simpler 1s system, plus that of beryllium, the second s^2 system, augmented by — i.e., corrected for — effects of an added inner shell:

$$He \approx H + [Be + (Be - Mg)] = \textbf{24.596}$$

Appendix IV

Integers, Elements, and Mendeleev's Line

How — and how far — people count tells a lot about their culture. Counting by mathematicians goes "1 2 3 4 5 6 ..."; by organic chemists "methane, ethane, propane, butane, pentane, hexane ..."; by inorganic chemists "hydrogen, helium, lithium, beryllium, boron, carbon ..."; and by spectroscopists

"K L M N . . .", in the x-ray region of the electromagnetic spectrum, and "*s p d f g h* . . .", in the uv and visible regions.

All counting schemes are based on an atomic model of matter; i.e., on the existence of discrete, countable entities. To physical science's list of seven basic physical quantities — mass, length, time, absolute temperature, electric current, luminous intensity, and amount of substance — counting contributes one quantity: "amount of substance", symbol *n*, meaning "population of elementary entities".

In the counting of organic chemists the counted entities are CH_2 groups in a homologous series of molecules. In the counting of inorganic chemists, the counted entities are protons in atomic nuclei. In both instances nature generates, in words of Bertrand Russell, speaking of the natural numbers, "A series in which there is a first term, a successor to each term (so that there is no last term), no repetitions, and every term can be reached from the start in a finite number of steps . . ." Such series, says Russell, are called *progressions*. "Progressions," he adds, "are of great importance in the principles of mathematics" — and also, one may add, in the physical sciences. "Any progression," continues Russell, "may be taken as the basis of pure mathematics." The progression of the chemical elements — called at the outset of this report "Mendeleev's Line" — is the basis of inorganic chemistry.

Since one never knows in mathematics what one is talking about, observed Russell, the terms of its fundamental progression may be any *things*. Usually they're called in mathematics the natural numbers. In the early history of the Periodic System inorganic chemistry's fundamental progression (Mendeleev's Line) was often compared with the homologous series of organic chemistry, to no avail. The relation of increments along a progression to the progression's first member is different in the two instances. For homologous series, increments are CH_2. The first member of the paraffinic series is, however, CH_4. For the chemical element progression, on the other hand, essential increments in atomic nuclei are the proton. And the nucleus of the progression's first member contains a single proton.

The character of the chemical element progression lies closer to that of Peano's fifth axiom for arithmetic than does the character of organic chemistry's homologous series. Peano's fifth axiom states that any property which belongs to the first number and belongs to the successor of every number that has that property, belongs to all numbers. It is the basis of mathematical inductions. The corresponding features in the history of chemical thought were Mendeleev's inductions of missing elements, based on the assumption that there are no gaps in his line. The analogy

Mendeleev's Line-Up of the Elements stands to Inorganic Chemistry as the Natural Numbers stand to Mathematics

embraces two fundamental questions. In mathematics: *"What is a number?"* In chemistry: *"What is an element?"* Definition of number was given by Frege in 1884, writes Russell, in terms of the concept collection, aggregate, manifold, class, or set, defined in two ways, notes Russell: by "extension", i.e., by enumeration — impractical for large sets; and by "intension", i.e., by some distinctive characteristic.

"It is [the] fact that a defining characteristic is never unique," says Russell, "that makes classes useful." The halogens may be defined by their atoms' atomic numbers, for example, or by numbers and types of valence-shell electrons. Or they may be defined by their salt-generating character as simple substances. Russell does not discuss how to judge the usefulness of different classifications of the same entities based on different defining characteristics.

The fleas come with the dog. The fundamental reason for the usefulness of classes is the fundamental reason for existence, in chemistry, of The Helium Problem.

Which is more useful: Definition of the Alkaline Earth Metals as elements whose atomic numbers are generated by the sequence of integers 1 1 2 2 3 3 4 4 squared, doubled, and then added to each other (**3**); and definition of the Noble Gases as the previous numbers minus 2 — with, in both instances, first members of Groups not members of triads? Those definitions place helium above beryllium in periodic tables. Or is it more useful to define the Alkaline Earth Metals as divalent, alkali-forming metals and the Noble Gases as noble gases? Those definitions place helium above neon in periodic tables.

Definitions in science are elective, but not arbitrary. They are created in strict agreement with Nature's nature. The answer to the question Which definitions of Alkaline Earth Metals and Noble Gases are the most useful is, of course, another question: Useful in what context? In the context, e.g., of a *rule of similarity and adjacency* of simple substances in periodic tables? (That rule has a number of striking exceptions involving first-row elements, particularly in the *p*- and *s*-blocks.) Or in the context of a *rule of first-element distinctiveness* of simple substances? (That rule has few exceptions. Leading ones appear to involve heavy elements for which relativistic effects may be important, particularly in the case of mercury in the Volatile Metals Group Zn Cd Hg. Mercury, compared to cadmium, is, in many respects, more distinctive than is its Group's first element, zinc, compared to cadmium.) An answer: First things first. First First-Element Distinctiveness.

In hindsight, issues of leading importance regarding fundamental features of the Periodic System appear first in this report. Although it may have seemed at the outset elementary and scarcely worth mentioning, *The Helium Question*, cited at the outset [and returned to "again and again", albeit deemed by one reviewer of an early version of a part of this report to raise a question where none exists (2)], and *Mendeleev's Line* [pictured on the report's front cover, albeit seldom cited in modern textbooks of general and inorganic chemistry] focus attention, jointly, on the essence of the Periodic System: existence of a physical *progression,* in the strictest mathematical sense of that term, and different Group-forming definitions, in terms, e.g., of atomic numbers (and the Triad Rules), atomic structure (and numbers and types of valence-shell electrons), or chemical and physical properties of atomic aggregates (called by Mendeleev "simple bodies").

Classification of the elements by the Periodic Law and properties of simple bodies, historically the first method, is the least reliable, for light elements, owing to the Phenomenon of Light-Element Distinctiveness. Classification by atomic structure, the common practice, today (except for helium!), encounters numerous difficulties in the *d*- and *f*-blocks. Only classification according to a set of natural table construction conventions is free of ambiguities.

Appendix V

Mendeleev's Remarks Regarding Artificial and Natural
Classifications of the Chemical Elements (13)

With added *emphasis* [and editorial remarks]

"All systems [for classifying the chemical elements] known up to the present can be placed in two distinct categories.

" Into one of these categories *(artificial systems)* enter the systems which are **based on some few characters of elements.** For example, the systems of division of the elements according to their affinity [or its absence], according to their electro-chemical properties, according to their physical properties [low boiling points, e.g.] *(or division into metals and metalloids)*, according to the manner in which they behave in the presence of oxygen and of hydrogen, according to their atomicity, &c. These systems, in spite of their evident defects [in the case of helium, when grouped with the other noble gases: violation of the Rules of Triads, Group Sizes, First-Element Distinctiveness, Spectroscopic Classification of Atoms, and Atomic Structure] deserve some consideration, for they have the merit of a certain exactness, and each of them has contributed in a special direction to the progressive elaboration of chemical ideas [regarding, e.g., the chemical significance of subshell closure and the principal quantum number n].

"The systems of the second category (natural systems) establish, according to members of different and purely chemical properties, groups of analogous elements.

"[The Periodic Law] gives us the means to erect an unarbitrary system, as complete as possible [based on two atomic properties:] the *forms of oxides* [their chemical formulas; i.e., their _atomic ratios_] and . . . _atomic weights_."

Appendix VI

The Case for Locating Hydrogen above Fluorine
in Periodic Tables

By the Theory of Substitution, hydrogen belongs above fluorine in periodic tables. The two elements form the same types of compounds. And the compounds have the same types of chemical formulas: e.g., NaH and NaF, CH_4 and CF_4.

Broadly speaking, three types of substances exist: Covalent, Ionic, and Metallic (C, I, and M). Usual electrical conductivities in the solid and liquid (molten) states are, respectively (P = Poor, G = Good): P&P, P&G, and G&G.

Type of Substance:	Covalent	Ionic	Metallic
Solid State Conductivity	P	P	G
Liquid State Conductivity	P	G	G

By those criteria, atoms of the cited elements, H and Na, C and F, form with each other the following types of substances.

	Na	C	F	H
Na	M	I	I	I
F	I	C	C	C
C	I	C	C	C
H	I	C	C	C

For over three decades that matrix and the Theory of Substitution convinced the author that hydrogen's natural location in periodic tables was above fluorine. As the matrix shows, no substances remain of the same type when sodium is substituted for hydrogen. The H and Na columns and rows are, so to speak, orthogonal to each other.

Appendix VII

Distinctive Features of the *s*-Block

In all Periodic Tables

- Fewest number of Groups

- Two full Triads per Group (for Z through 120)

- Site of Dobereiner's first Triad (Ca Sr Ba)

- Common names for all Groups

- Site of the most reactive metals

- Site of six elements with the lowest first stage ionization energies

- Mendeleevian numerals (I, II) correspond to Groups' ordinal numbers within the block (In other blocks the former is the latter plus 2)

- Smallest maximum Group oxidation number (+1)

- Smallest largest maximum Group oxidation number (+2)

- Second Group's valence electrons remain such, with increasing Z, through *f*-, *d*-, and *p*-blocks

- Its elements are not involved in secondary kinships

- One element is involved in a "special kinship": Mg with Zn

- Site, on entrances to the block from the left, of chemistry's most striking changes in elements' properties with increasing atomic numbers

- Site, with increasing Z, of occupancy of new electron shells

- Orbitals of differentiating electrons contain only radial nodal surfaces

- For orbitals of differentiating electrons, number of domains, defined by nodal surfaces, equals *r*, the number of radial nodal surfaces. (For other blocks the number of domains is $2r\ell$.)

In Tables with H and He in the s-Block

- Site of the largest Groups

- Site of the two lightest and two lowest boiling elements

- Site of the only elements that have no inner shell electrons

- Site of elements with the smallest and largest atomic cores

- Site of the least reactive element

- Site of the element with the highest first-stage ionization energy

- Site of the Periodic System's leading examples of First-Element Distinctiveness

- Sites of leading examples of Second- and Third-Element Distinctiveness

- Site of leading examples of tertiary kinships: H with F, He with Ne

- Period and row numbers, **P** and **r**, the same

In the Left-Step Table

- Contains all elements of the first Dyad

- Pivotal Block (where Periods end)

- Yields the Period Rule: $P = r + 2\ell$

- Yields the Row Rule: R(row ordinal number in Table) $= (P^2 + 2P + \delta)/4 - \ell$

- Origin or terminus of block-to-block trends

Appendix VIII

Broken Symmetry and Ruptured Regularity

A statement by Richard Feynman, forwarded by our friendly physicist, Robert Bent, suggests an analogy:

"True" Periodic Tables are to the Left-Step Table

as

True Planetary Orbits are to Perfect Circles

Writes Feynman (in *Six Not-So-Easy Pieces,* pp46-47):

> "We have, in our minds, a tendency to accept symmetry as some kind of perfection. In fact it is like the old idea of the Greeks that circles were perfect, and it was rather horrible to believe that the planetary orbits were not circles, but only nearly circles. The difference between being a circle and being nearly a circle is not a small difference, it is a fundamental change so far as the mind is concerned. There is a sign of perfection and symmetry in a circle that is not there the moment the circle is slightly off — that is the end of it — it is no longer symmetrical. . ."

Similarly, there is perfection and regularity in a block-style periodic table whose blocks are arranged by size that is not there the moment the arrangement is slightly off — that is the end of it — it is no longer perfectly regular.

Lost with the broken symmetry of a perfect circle is the constant ratio of circumference to diameter, π. Lost with the ruptured regularity of the perfectly regular LSPT is the constant sum along a period of the Madelung parameter, $n + \ell$. Gained in the former instance is an accurate account of the planet's orbits. Gained in the latter instance is, in Robert Frost's words, "the straight crookedness of a good walking stick": the straight irregularly of blocks' sizes in the runs through maximum states of oxidation along periods of a "true" periodic table.

"Perhaps broken symmetry [and sacrificed regularity] makes nature uglier than the ideal but much more interesting," writes Robert Bent's colleague Roger Newton (in *What Makes Nature Tick?), "*or perhaps it makes it even more beautiful because somewhat wacky . . ."

The *s*-block is "somewhat wacky" in that, for instance, it is the only block in the Periodic System with a Group of elements whose maximum state of oxidation is +1 and whose outer electrons (except in the case of helium) go on with increasing atomic number to be valence electrons in other blocks.

Beauty, one is reminded, resides in the eye and mind of the beholder. To users of periodic tables unmindful of the regularities of Periodicity displayed by the LSPT, the arrangement of metals and nonmetals in the Conventional Periodic Table is beautiful. To physicists unmindful of elements' maximum states of oxidation, the connection between the quantum numbers of Atomic Physics and the integers of Chemical Periodicity displayed by the LSPT is beautiful. To philosophers of science mindful of both kinds of regularities and irregularities, existence of *both* tables is beautiful. Their presence in a scientist's tool kit is like having a screwdriver for screws (the screwy *s*-block) and a hammer for nails (to nail down connections between chemical and physical thought).

Feynman ends with this thought: "[Some] think that the true explanation of the near symmetry of nature is this: that God made the laws only nearly symmetrical so that we should not be jealous of His perfection." Others may think that the true explanation for the need for more than one periodic table is this: that God made the initial conditions for the Big Bang such that evolution of the universe would lead to many-electron atoms and a wacky *s*-block, in order that chemists and physicists would not be bored.

Appendix IX

Alternative Punctuations

s sp sp sdp sdp sfdp sfdp   and   ss ps ps dps dps fdps fdps

According to U.S. Senate lore, following the bloody Missouri-Kansas fighting at the time of the Civil War, a Senator from Kansas reported to the Senate that parents of a Kansas girl who'd eloped with a boy from Missouri found this note on her bedroom dresser:

"Goodbye God [sfdp], I'm going to Missouri [fdps]."

With all due deference to his honorable colleague from Kansas, replied Missouri's famous Senator Bennett Champ Clark, what the girl had written was:

140

"Good! By God, I'm going to Missouri!"

It's possible to harbor both sentiments — regret, when using fdps periods, at the loss of the definitive period endings of the punctuation sfdp(eriod), a full stop before proceeding to occupancy of a new shell; and pleasure, in the regularities of the fdps punctuation — by dwelling, so to speak (Figure 9), in Kansas City, Missouri: sfdps.

Appendix X

Mendeleev's Absolute Distinction
and
Whewell's Fundamental Antithesis of Philosophy

Mendeleev's <u>absolute distinction</u> between *simple bodies* and *atoms* corresponds to Whewell's <u>antithesis</u> or <u>opposition</u> of *things* and *thoughts*, *sensations* and *ideas*, *facts* and *theories*, *nature* and *man-added-to-nature*.

"We have no knowledge without the union, no philosophy without the separation of the two elements," emphasized Whewell.

In the concrete case of helium and periodic tables, we have no knowledge (in a modern chemical sense) of "helium" without a union in our minds of the reason for its leading properties as a simple body — especially its inertness and volatility — in terms of its atomic structure. And we have no philosophy to guide us in locating its symbol in periodic tables without a separation in our minds of *those two "heliums"*: an inert gas and an atomic s^2 system — for which, unfortunately (as pointed out by Mendeleev), we haven't distinctive names and symbols.

> The situation is analogous to what happened after G. N. Lewis identified the valence stroke of classical bond diagrams as two electrons. What, then, did the symbols of the elements stand for? Atomic cores. In the simplest case, the bond diagram for a molecule of dihydrogen, H_2, may be written (with Lewis' "absolute distinction" in mind) **H—H**, where "—" stands for 2 electrons and "**H**" stands for H^+. Neither Mendeleev's "absolute distinction" nor Lewis' "absolute distinction" has been encoded in chemical notation. Produced, consequently, has been a "fixation of outlook" (Whewell) unreceptive to classification of helium with the alkaline earth metals — and covalent compounds as ion-compounds (71).

The object of any scientific Classification, notes Whewell, is enunciation of scientific truths. "[I]t must enable us to state true and general propositions" — such as a block-to-block trend in first-element distinctiveness (**31**); or an isomorphism between the rules of covalent compounds and Pauling's Rules of Crystal Chemistry (71). Success in both endeavors depends on holding in mind the distinction between a fact (helium is an inert gas; covalent compounds are nonpolar) and a theory (helium atoms are similar to atoms of the alkaline earth atoms; all compounds may be viewed as ion-compounds [in which cations = atomic cores and anions = electride-ions {corresponding to the valence-strokes of bond diagrams}]).

Historically speaking (to use Whewell's words and ideas regarding classification of plants) a *serious inconvenience* has been this: the task of classification in both of our cases (of helium and chemical compounds) was performed when the governing laws were not known. Consequently, selection of resemblances taken into account could be only those known at the time. "If these selected [resemblances] be then made absolute and imperative, and if we abandon all attempts to obtain Natural Classes of any higher order and wider extent," continues Whewell, "we form [and pass on to posterity] an Artificial System."

Appendix XI
Centrally Productive "Magic Formulas"

Leading achievements of human thought in mathematics and the physical sciences are often expressed algebraically. Famous examples from physics are Newton's formula f = ma and Einstein's equation $E = mc^2$.

In the theory of complex numbers "the principal analytical findings up to the middle of the 18th century," writes Salmon Bochner (70), "culminated in the 'magic' formula

$$e^{ix} = \cos x + i \sin x$$

"In this formula the letter x denotes arc length so that in particular

$$e^{i\pi} = -1, \quad e^{2i\pi} = 1$$

"The great strength of these relations stems from the fact that the three symbols

$$e, \ i, \ \pi$$

which they bring together, are, by *provenance*, unrelated to each other. Such linkages of three mathematical objects of disparate provenance in one formula are exceptional; in general, even centrally productive formulas link up only two separate mathematical objects at the same time."

The following centrally productive formula in thermodynamics brings together *four* separate *physical* quantities at the same time:

$$(\partial E/\partial S)_V = T$$

The three symbols on the left, E, S, V, are the primary extensive physical quantities of thermodynamics. T is its leading intensive physical quantity. The great strength of the relation stems from the fact that four quantities that it brings together are, by *provenance*, unrelated to each other.

A centrally productive formula from a union of chemistry and physics that brings together four separate symbols is

$$P = r + (\Delta h)\ell$$

It implies that (cf. Table 5):

$$(\Delta r/\Delta \ell)_P = -\Delta h$$

The great strength of the relation stems from the fact that its four symbols

$$r, \ell, P, \Delta h$$

have *two sets of meanings* of disparate provenance: the provenance of *chemical periodicity* where ℓ = block number, r = a block's row number, P = period number in the Left-Step Table, and Δh = step height, 2); and the provenance of *atomic physics* where ℓ = angular quantum number, r = radial quantum number, P = order of orbital occupancy parameter for many-electron atoms, and Δh = a weighting factor, depending on the magnitude of effects on orbital energies of electron screening and penetration.

Appendix XII

Positive and Negative Features of sfdp and fdps Periods

+ positive   ⟶ very positive   • neutral   - negative

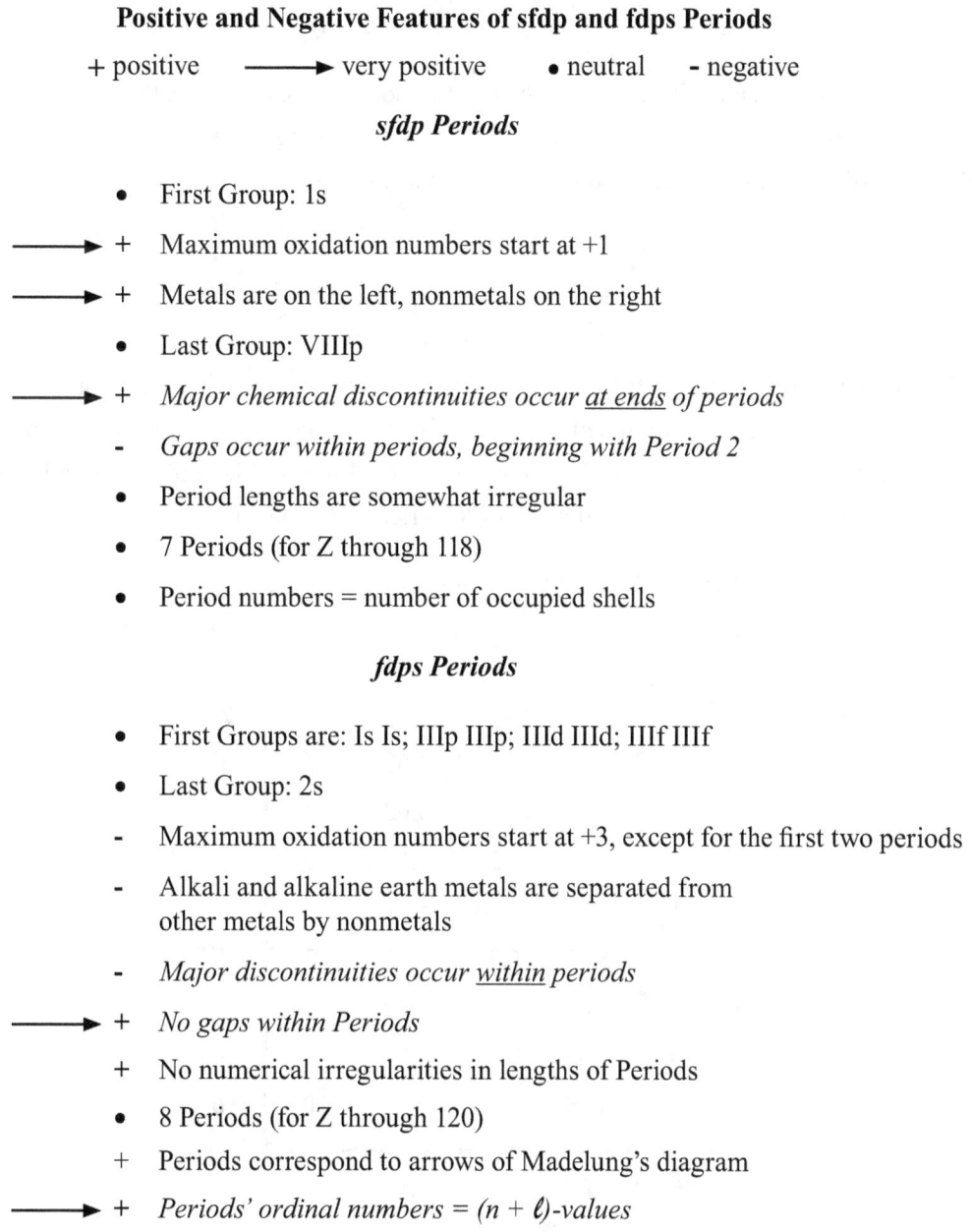

sfdp Periods

- •   First Group: 1s
- ⟶ +   Maximum oxidation numbers start at +1
- ⟶ +   Metals are on the left, nonmetals on the right
- •   Last Group: VIIIp
- ⟶ +   *Major chemical discontinuities occur <u>at ends</u> of periods*
- -   *Gaps occur within periods, beginning with Period 2*
- •   Period lengths are somewhat irregular
- •   7 Periods (for Z through 118)
- •   Period numbers = number of occupied shells

fdps Periods

- •   First Groups are: Is Is; IIIp IIIp; IIId IIId; IIIf IIIf
- •   Last Group: 2s
- -   Maximum oxidation numbers start at +3, except for the first two periods
- -   Alkali and alkaline earth metals are separated from other metals by nonmetals
- -   *Major discontinuities occur <u>within</u> periods*
- ⟶ +   *No gaps within Periods*
- +   No numerical irregularities in lengths of Periods
- •   8 Periods (for Z through 120)
- +   Periods correspond to arrows of Madelung's diagram
- ⟶ +   *Periods' ordinal numbers = (n + ℓ)-values*
- ⟶ +   *Periods suggest, in concert, locations for "problem elements"*

Appendix XIII

Fibonacci Character of Chemical Periodicity

$$u_{n+2} = u_n + u_{n+1} \quad n = 0, 1, 2, \dots \quad u_0 = 0 \quad u_1 = 1$$

$$0 \quad 1 \quad 1 \quad 2 \quad 3 \quad 5 \quad 8 \quad 13 \quad 21 \quad 34 \quad 55 \quad 89 \quad 144 \quad 233 \ \dots$$

The next member in the series r, $r + \ell$ ($= n$), $r + 2\ell$ ($= P$) might seem to be $r + 3\ell$. Taken as a description of periods' ordinal numbers P, P $= r + 3\ell$ yields triadic tables, which, in chemistry, are of only theoretical interest, in providing perspective on Periodicity's dyadic character (Figure 27). Missing from the previous series r, $r + \ell$, $r + 2\ell$ is, near the outset, the term ℓ.

$$r \quad \ell \quad r + \ell \quad r + 2\ell$$

The series' third term, $r + \ell$, is the sum of the two previous terms. Likewise, the series' fourth term is the sum of its two previous terms. The sequence is the beginning of a series of linear forms in r and ℓ whose coefficients are adjacent terms in the Fibonacci series -

$$
\begin{array}{ll}
u_0 = r & u_4 = 2r + 3\ell \\
u_1 = \ell & u_5 = 3r + 5\ell \\
u_2 = r + \ell & u_6 = 5r + 8\ell \\
u_3 = r + 2\ell & u_7 = 8r + 13\ell
\end{array}
$$

The order of u's numerical values for different values of r and ℓ depends on the ratio of the coefficients of r and ℓ. For higher and higher terms, ℓ's coefficient divided by r's coefficient approaches the numerical value $(1 + \sqrt{5})/2 = 1.618055$. For, e.g., terms 1 to 8 and 12, and 13, the ratio's values are: ∞, 1, 2, 1.5, 1.67, 1.6, 1.625, 1.615, 1.61798, 1.6180 $\approx$ 1.62. Cited below are -

Numerical Values of r + 1.62 ℓ in the Format of the Left-Step Periodic Table

g	f	d	p	s		P
				1	1	
				2	2	
			2.62	3	3	
			3.62	4	4	
		4.24	4.62	5	5	P
		5.24	5.62	6	6	
	5.85	6.24	6.62	7	7	
	6.85	7.24	7.62	8	8	
7.47	7.85	8.24	8.62	9	9	
8.47	8.85	9.24	9.62	10	10	

144

Numerical values of the Periodicity parameter r + k ℓ , k = 1.62 increase monotonically (left-to-right, top-down in the table above) for its first 12 entries, through 6 (in period **6)**. Values of the Fibonacci order-of-occupancy parameter for ns orbitals are, for $n \geq 4$, close to (and slightly less than) those for (n-1)d orbitals. For n $\geq$ 6, the ns and (n-2)f orbitals order of occupancy parameters are nearly the same.

Broadly speaking, Periodicity exhibits Fibonacci character, in the sense that the ratio of the ℓ-dependence to the *r*-dependence of orbital energies of many-electron atoms is generated, approximately, by a limiting ratio of a series in which a term's magnitude depends on the magnitudes of the *two previous terms*, which corresponds to the fact that, owing to screening effects of inner electrons, an orbital's energy depends on the electron density created by previously added electrons.

The value of 7.47 for the 5g orbital ($n + \ell = 9$) lies below the Fibonacci values for the orbitals 8s and 7p ($n + \ell = 8$) and close to the 7.24 of the 6d orbital (n + ℓ = 8). And, indeed, for potassium atoms the 5g level does lie below the 8s, 7p, and 6d levels (Figure 28).

Appendix XIV

Derivation of an Expression for Row/Orbital Ordinal Numbers R

R-values for an ℓ-column of rows (e.g., for $\ell = 2$, R = 7, 10, 14, 18) are equal to the corresponding R-values for the $\ell = 0$ column (9, 12, 16, 20), less ℓ.

R-Values, fdps Format

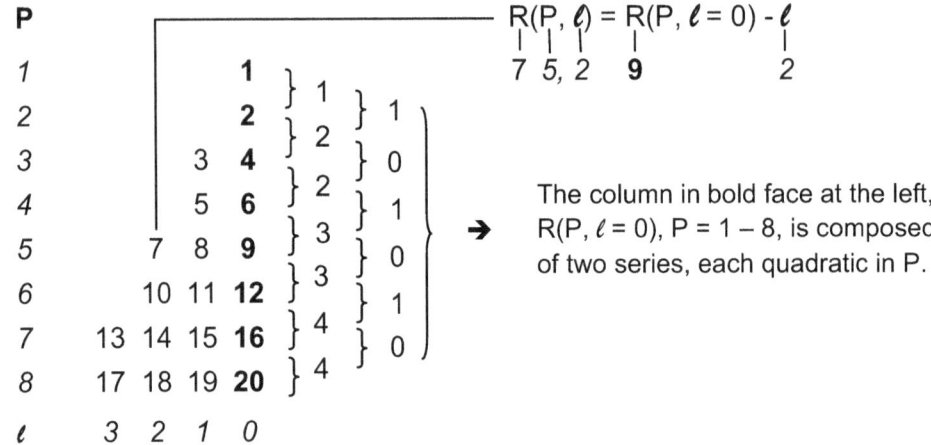

The column in bold face at the left, R(P, ℓ = 0), P = 1 – 8, is composed of two series, each quadratic in P.

R-values for the $\ell = 0$ column for odd P-values are the square numbers, starting at 1.

P_{odd}(= 2n -1)	n	R(P, ℓ = 0)
1	1	**1**
3	2	**4**
5	3	**9**
7	4	**16**

By inspection, at left

$$R(P, \ell = 0) = n^2 = [(P + 1)/2]^2$$
$$= (P^2 + 2P + 1)/4$$

R-values for the $\ell = 0$ column for even P-values are twice the sums of the integers, starting at 1.

P_{even}	R(P, ℓ = 0)	R/2	P/2
2	**2**	1	1
4	**6**	3	2
6	**12**	6	3
8	**20**	10	4

$$R(P, \ell = 0) = \sum (P/2)$$
$$= (P/2)[(P/2) + 1]/2$$
$$= (P^2 + 2P)/4$$

Summary: $R(P_{odd}, \ell = 0) = (P^2 + 2P + 1)/4$ $R(P_{even}, \ell = 0) = (P^2 + 2P)/4$

➔ $R(P_{even\ or\ odd}, \ell = 0) = (P^2 + 2P + \delta)/4$ where δ = 0 or 1 as P is even or odd

➔
$$\boxed{R(P, \ell) = (P^2 + 2P + \delta)/4 - \ell}$$

146

Appendix XV
Desirable Features of Periodic Tables

Filled circles (•) designate features that require He/Be.

Italicized features in the first two categories are possessed, also, by the LSPT.

Features Possessed by the Conventional Periodic Table: He/Ne

o *Arrangement of elements by atomic numbers*

o *Vertical display of primary kinships*

o *Gapless columns*

o *Periods' rows in horizontal arrays*

o *Nonmetals grouped together* (except for H)

o Metals grouped together

o Graphic display of the striking discontinuities in elements' properties on entering the *s*-block

o Horizontal trends in electronegativity, metals to nonmetals

o Periods begin with Group I

o Graphic display of numbers of occupied subshells

Features Possessed by the Conventional Periodic Table: He/Be

• *Placement of elements as atoms, not as simple substances*

• *Adherence to descriptions by atomic physics of atoms' ground states*

• *Adherence to Triad Rules*

• *Adherence to a Rule of Group Sizes*

• *Adherence to a Row-Orbital Correspondence Principle*

Features Possessed by the Left-Step Periodic Table

Features italicized above

• Easy construction from the Periodic Law

• No irregularities

• No "floating" or "submerged" blocks

• Easy description of the Table's shape

• Easy stipulation of atoms' electronic structures

• Exhibition of two types of Main Groups, by size

• Exhibition of four types of Groups, by size

• Graphic display of Periodicity's dyadic character

• Graphic display of the Madelung's parameter $n + \ell = r + 2\ell$

• Suggestion of a physical explanation for periodic tables' shapes

• Graphic display of the Filled-Shell Rule

• Dyad numbers

• Unambiguous locations for the "problem elements"

o   Unique locations for all elements

o   No elements in special "series" rather than in periods

o   Gapless periods

o   Blocks ordered by size

o   Support of searches for and graphic displays of block-to-block trends

o   Period ordinal numbers for graphic display of secondary periodicity

o   Gases together

o   Main Groups together

o   Commonly named Groups together

o   Suggestion of systems of natural column labels

o   Open-ended, for ease of addition of new elements

<u>Other Features</u>

o   Graphic display of secondary kinships

    "Short-Form" Tables
    Step-Pyramid Tables
    Front-Step Tables

•   Graphic display of tertiary kinships

    Step-Pyramid Table, s-elements on the left

o   Graphic display of the special zinc-magnesium kinship

    Step-Pyramid Tables

o   Graphic display of classical transition elements

    "Short-Form" Tables
    Step-Pyramid Tables
    Front-Step Tables

o   Graphic display of blocks' rows' ordinal numbers in the Periodic System

    Right-Step Table
    Modern "Short-Form" Table
    "Color-Numeric" Table

•   Display of numbers and types of valence-shell electrons along different coordinates

Appendix XVI

The Step-Pyramid Table, *s*-Elements on the Left

The sfdp Step-Pyramid Periodic Table is unrivaled in the ease of graphic representation of non-primary kinships. Indicated in Figure 53 are Periodicity's leading secondary kinships, between second-row elements of the *p*- or *d*-blocks and first-row elements of the *d*- or *f*-block, respectively. Not indicated are strong secondary kinships between Th and Hf, Pa and Ta, and U and W. Compromised by the table's appealing, symmetrical shape is congener verticality. Irregular are slopes of primary kinship lines. For *s*-elements they slope downward to the left, for other elements downward to the right. Highlighted is Periodicity's dyadic character, for periods 2 – 7. The top period consists of the arithmetically, if not chemically, irregular hydrogen-helium monad. It may be viewed in two complementary ways.

Two Views of the sfdp Step-Pyramid's H-He Monad

1. The monad is so distinctive it warrants a special definition (provided by Jensen) of primary kinships that denies H and He primary kinships with other elements.

2. The monad is striking graphic representation of the Periodic System's leading example of the First-Element Distinctiveness Rule (of Mendeleev and Jensen).

Either way, one follows in the footsteps of Professor Jensen. Called to mind is the ever-present figure sketched peering over bulkheads during WWII above the inscription: "Kilroy was here."

The situation is analogous, in some ways, to two interpretations of the observation around the turn of the last century of the independence of the properties of cathode rays on the chemical nature of the cathode-ray tubes' cathodes and residual gas. The rays have something to do, proposed a German school of physicists, with a feature common to all the experiments: the "ether". They are a universal constituent of matter, proposed J. J. Thomson. In both instances, a choice between two interpretations of the same thing is made in the same way: Which interpretation is most useful? Of course, no analogy is perfect. (Otherwise the analogue would be the analogued thing itself.) In the case of the hydrogen-helium monad, both views are useful.

Because of the extraordinary character of H and He, it's pleasing to see a useful and visually attractive periodic table assign the two elements in a natural manner to an extraordinary location. Similarly, it's pleasing to see a useful and irregular-free periodic table assign H and He in a natural manner to a non-extraordinary location as an extraordinary instance of the Mendeleev-Jensen Rule. Both views agree on one thing: Hydrogen and helium are — arguably — the most extraordinary elements in the Periodic System.

Appendix XVII

What Are Periodic Tables and What Are They Used For?

Periodic Tables are tabular expressions of the Periodic Law: "The elements, if arranged according to their atomic weights, exhibit an evident *periodicity* of properties" (Mendeleev, 1869). Elements' symbols appear in most periodic tables in vertical columns, called Groups, and, simultaneously, in horizontal rows, called Periods. The result is the "Periodic Classification of the Elements" according to Groups and Periods. To understand its distinctive character, it's important to know, emphasized Mendeleev, what it is not. It is not a classification of the elements as ordinary, simple substances, such as graphite, ozone, or inert gases. It's a classification of *atoms*, based, in the first instance, in Mendeleev's work, in the days before electrons and orbitals, on *atomic* weights and *atomic* combining capacities, determined by maximum states of oxidation. Early periodic tables were often called, accordingly, "Oxidation Tables" or "Valence Tables". "Periodicity" occurs along Mendeleev's lineup of elements according to their atomic weights (today: atomic numbers) in the form of similar sequences, lengths 2, 6, 10, and 14, in the pattern

$$2\,2 \quad 6\,2 \quad 6\,2 \quad 10\,6\,2 \quad 10\,6\,2 \quad 14\,10\,6\,2 \quad 14\,10\,6\,2$$

In many periodic tables, Groups appear, accordingly, in blocks of 2, 6, 10, or 14 Groups, called the *"s"*, *"p"*, *"d"*, and *"f"* or the $\ell = 0, 1, 2,$ and 3 blocks. Illustrated is the leading purpose of periodic tables: exhibition of regularities that are latent in Mendeleev's Line but that are not immediately apparent to the unaided eye, particularly *chemical kinships:* primarily primary kinships [elements in the same Group (usually indicated, as said, by verticality in periodic tables)]; secondarily secondary kinships [elements in different blocks, hence in different Groups (indicated, in different types of periodic tables, by tie-lines, adjacency, or alphanumeric Group labels) that have the same *maximum* states of oxidation and that are located the same distances from their blocks' *left* ends]; and, thirdly, tertiary kinships [elements in different Groups that have the same *minimum* states of oxidation and that are located the same distances from blocks' *right* ends]. The helium-neon kinship, like the hydrogen-fluorine kinship, is a tertiary, not a primary, kinship. [Placement in periodic tables of He, an s2 atom, above Ne, a p6 atom, is a (usually unique) category mistake: two different categories of classification (elements as *atoms* and elements as *simple substances*, such as "inert gases") united in a single classification.] Whatever their arrangements of Groups, all periodic tables are topologically equivalent string-figure variations of a Mendeleevian lineup of the elements. Most periodic tables are read, for increasing atomic numbers, Z, left-to-right, top-down. Elements' symbols in the same column (whose entries therefore occur periodically in passage through the tables with increasing Z) stand for atoms that are generally alike regarding numbers and type(s) of valence-shell electrons *(s, p, d, or f)*. Also, the corresponding simple substances are generally similar to each other, *except in the case of elements of small atomic number* (especially H through F), which, compared to their congeners, have highly distinctive properties, *particularly in the case of Nature's two lightest elements, hydrogen and helium,* provided they appear in periodic tables above lithium and beryllium, respectively. Then, illustrated *throughout* periodic tables is one of the leading achievements of scientific thought: a correspondence between the ordinal numbers generated by periodic tables and the leading quantum numbers of atomic physics. Blocks' ordinal numbers, ℓ, starting at 0 for the block of 2 Groups and ending at 3 for the block of 14 Groups, and blocks' rows' ordinal numbers within blocks, r, starting at 1, are equal, respectively, to the angular and radial quantum numbers of the orbitals occupied by the blocks' preponderant type of differentiating electrons. Lengths of blocks' rows (the "element-sequences" spoken of at the outset) are $2(2\ell + 1)$. Major changes in elements' properties with increasing Z occur on entrances to the *s*-block, a particularly distinctive block, with first appearance there of new values of the index of leading discontinuities, and principal quantum number, $n = r + \ell$. Order of occupancy with increasing atomic number of blocks' rows is always given, and for the corresponding atomic orbitals is usually given, by the empirical rule: smallest values of $r + 2\ell = n + \ell$ first and, for given $n + \ell$, largest ℓ first. Ordinal numbers of the periods of the Left-Step Periodic Table, in which blocks occur in the order *f d p s*, are equal to the values along its periods of $n + \ell$. Blocks' order in the Conventional Periodic Table is *s f*(usually footnoted) *d p*. In that sequence of blocks, runs through maximum states of oxidation along periods start at I. Proceeding onward, to the right, metallic character, reducing strength, and oxide basicity decrease (provided one ignores the Noble Gas Group and irregularities in the *d*-block). In all Groups metallic character increases downward, particularly quickly, and markedly, in the s-block (with hydrogen above lithium and helium above beryllium), less quickly in the *p*-block, and, continuing on to higher *ℓ*-blocks, still less quickly in the *d*- and *f*-blocks, all of whose elements are metals. A number of properties of first-row elements, in addition to their distinctiveness with respect to their heavier congeners, exhibit a similar *block-to-block trend* with increasing or decreasing values of *ℓ*. The trends are destroyed, however, when helium appears above neon and, otherwise, are hidden from view when the *ℓ* = 3 block is footnoted and the remaining three blocks appear in the order *ℓ* = 0 2 1. Exhibition of block-to-block trends requires periodic tables whose block order is fdps (or spdf). On the other hand, exhibition of conventional, horizontal trends in, e.g., metallic character (cited above) requires periodic tables whose block order is sfdp. Illustrated is the fact that no periodic table is superior to all other periodic tables in all respects. Periodic tables' intrinsic shapes stem from late occupancy in Bohr's Aufbau Process of large-*ℓ* orbitals, owing to effects of screening from atomic nuclei by inner-shell electrons of electrons in non-penetrating (large-*ℓ*) orbitals. An orbital's quantum number *r* is equal to its number of radial nodal surfaces, counting the one at infinity. It's quantum number *ℓ* is equal to (i) the number of *nucleus-containing* nodal surfaces and, also, to (ii) the power of a *nodal point at the nucleus*. Owing to those two facts, large-*ℓ* orbitals are "non-penetrating", relatively high-energy orbitals in many-electron atoms, the more so the larger *ℓ* and the larger the number of electrons. Use of the third dimension allows separation of the alphanumeric Group labels of two-dimensional periodic tables into their alphabetical and numerical components. Gained with that clarity are compact arrangements of atoms. Lost is graphic expression of Periodicity's periods. Illustrated, again, is the virtue of having at hand, or in the mind's eye, more than one tabular expression of the Periodic Law.

Appendix XVIII

Pauli Mechanics

For starters, in construction of chemical models of metals, replace (in the mind's eye) oxide ions of calcium oxide by "electride ions", e_2^{-2}. Or use the isoelectronic principle and move (alchemically) protons of the hydride ions of potassium hydride into the potassium nuclei. Produced, in both instances, is a rock-salt-like model of calcium metal, bonded by six-center, two-electron (6c/2e) bonds.

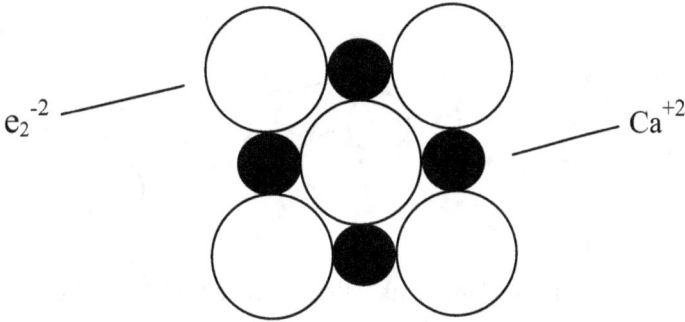

Centers of the calcium-core cations form a cubic-close-packed lattice. The electride-ion anions are powerful Bronsted bases, easily twice-protonated by water molecules to dihydrogen gas. Left behind are hydroxide ions, which precipitate out of solution with calcium ions, as slaked lime. (CAUTION! The reaction is highly exothermic! Water in a test tube containing a turning or two of "calcium electride" may boil!)

The electride ion model of bonding suggests images of the three types of chemical bonds: metallic (large cores only); ionic (large and small cores); and covalent (small cores only). Broadly speaking, there are no other possibilities.

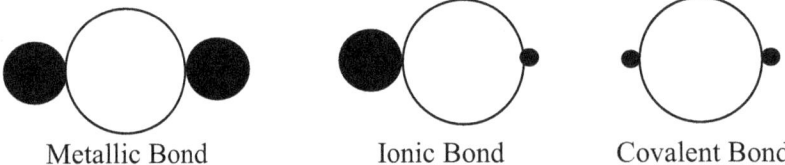

Metallic Bond   Ionic Bond   Covalent Bond

The dividing line between "large" and "small" occurs at a core radius of approximately half an angstrom. Since core radii increase downward and decrease rightward within blocks, a line of constant radii slopes downward to the right — the *p*-block's well known metalloid band.

To continue the previous line of thought, replace (with, e.g., small cork spheres) valence-strokes of valence-bond diagrams with space-filling, spherical electron-domains (69). Produced are ion-packing models of covalent compounds (71). With a "starter kit" of a tetrahedral set, two trigonal sets, a digonal set, and a single sphere, one can model the major lobes of common hybrid orbitals: s, sp, sp^2, sp^3, sp^3d, sp^3d^2.

Depicted above are electron-domain, highly schematic localized molecular orbital models of the valence-shells of the central atoms of H^-, BeH_2, BH_3, CH_4, PF_5 and SF_6. With the "starter kit" one can also construct electron domain models of the valence-shells of molecules and ions isoelectronic with ethane, ethylene, and acetylene (and, indeed, with sufficient spheres, models of all structures that have satisfactory bond diagrams, including ones that have — like "calcium electride" — multicenter bonds).

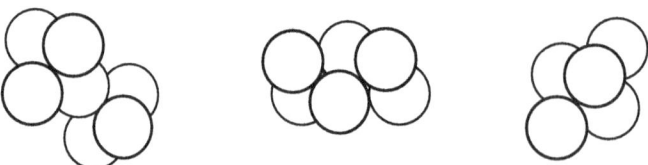

The three figures above — isomorphic with valence-stick models of C_2H_6, C_2H_4, and C_2H_2 — are copied from Pauling's Figure 13-21,—"The sharing of a corner, an edge, and a face by a pair of tetrahedra" in his section on "The Principles Determining the Structure of Complex Ionic Crystals" in his book *The Nature of the Chemical Bond*. Illustrated is a remarkably little-remarked-upon correspondence between the anion/cation packing model of ionic compounds and valence-stroke diagrams of covalent compounds (71). That correspondence, and conclusions cited in this report regarding Chemical Periodicity, bring to mind a Pennsylvania Dutch saying: "We're too soon old and too late smart." Had he been smarter sooner, the author might have spent his classroom chemistry-teaching years (1952-1992) teaching differently, Appendix XIX.

Appendix XIX
Suggestions for Chemistry Teachers from an Old-Timer

To Improve Education in (*and through*) Chemistry -

REFORM and EXPAND discussion of **Periodicity**.

> Replace He/Ne in periodic tables by He/Be.

> Expose students to distinctive features of several periodic tables.

Of course, to do so in a crowded curriculum, something must "give". Hence the suggestion:

STREAMLINE discussion of –

> **Molecular Structure**. Replace Lewis electron-dot formulas, Valence-Shell-Electron-Pair-Repulsion Theory, hybrid orbitals, and Molecular Orbital Theory with Pauling's Rules (of ionic crystals) applied (also) to bond diagrams of covalent compounds, via two substitutions: cations = atomic cores (represented in bond diagrams by elements' symbols); anions = electron-pairs (represented in bond diagrams by valence-strokes):

etc.

> **Quantity Calculus.** Replace the "factor label method", rhetorical algebra, and chapters on "The Mole Concept" with the relations -

$$d = m/V \quad M = m/n \quad c = n/V \text{ and}$$
$$N_A = N/n = 1/e = 12/\text{dozen} = \text{mol/mole}$$
$$\rightarrow \text{ mole} = \text{mol e, where mol} \approx 6 \times 10^{23}$$

Thermodynamics. Replace $\Delta U = Q + W$, "$1/T$ is an integrating factor for the reversible heat", and most of the rest of the conventional machinery of classical thermodynamics with -

$$E_{total} = E_{mech\ surr} + E_{thermal\ surr} + E_{system}$$
$$S_{total} = S_{mech\ surr} + S_{thermal\ surr} + S_{system}$$
$$\text{where:} \quad \Delta E_{th\ surr} = -\Delta E_{mech\ surr} \quad (\Delta E_{sys} = 0)$$
$$\Delta E_{sys} = -(\Delta E_{th\ surr} + \Delta E_{mech\ surr})$$

$$\Delta S_{mech\ surr} = 0$$

$$\Delta S_{th\ surr} = \Delta E_{th\ surr}/T$$

$$\Delta S_{sys} = -\Delta S_{th\ surr} \quad (\Delta S_{total} = 0)$$

DO **Demonstration-Experiments** (72).

Introduce no term or concept unless there is a need for it based on an observation or an experiment (73).

Exhibit oneself in one's own true trade (74).

Refer to periodic tables. Draw molecular structures. Exhibit molecular models. Propose reaction mechanisms. Explain where and in what manner energy is conserved and entropy increases.

MENTOR **Student Apprentices** learning chemistry (public speaking, team work, and responsible behavior, while feeling all the while that "this is the best thing that has happened to me in school") by doing and explaining demonstration-experiments for peers, younger students, and the general public.

USE demonstration-experiments as subjects for **Writing Assignments** (75).

Describe in simple declarative sentences that follow each other logically what was used, done, seen, said, written, drawn, imagined, and done next.

HAVE FUN with the atomic model of matter!

Appendix XX

Block-to-Block Trends

Four obvious trends are listed first.
Chemically leading trends follow, in ***bold face italics***.
Physically important trends come next, in *italics*.

Periodicity's two best-known trends are exhibited by essentially all periodic tables: a table-wide trend in atomic numbers; and a Group-wide trend, downward, toward increasing metallic character. Widely recognized, also, are period-wide trends, rightward, in sfdp tables, toward increasing nonmetallic character. Previously unrecognized are the block-to-block trends listed below (except for the first one).

Most of the inter-block trends involve first-row elements. Most of them require location of helium above beryllium. All of them require that the blocks be taken, mentally, if not so displayed, physically, in the order of their widths (either way).

- Block width (**91**).
- Group size (**91**).
- Ordinal numbers within the Periodic System of blocks' first-rows (**91**).
- Zs of blocks' last first-row elements (**90**).

- *Magnitude within Groups of First-Element Distinctiveness* (**31**).
- *Similarities of blocks' first-row elements to each other* (**91**).
- *First-stage ionization energies of blocks' first-row elements* (**91**).
  - o *Range of values*
  - o *Average values*
- *Rapidity of onset within Groups, downward, of metallic character* (**91**).
- *"ℓ-Nobility"* (**90**).
- *Maximum occupancy of valence-shells of first-row elements* (**91**).

- *Power of nodal points at atomic nuclei of the orbitals of blocks' predominant type of differentiating electrons* (**62**).
- *Number of nucleus-containing nodal planes of orbitals of blocks' predominant type of differentiating electrons* (**48**).
- *Number of circles and diameters in circle-diameter diagrams of atomic orbitals* (**71**).

- Odd-Row Distinctiveness (**60**).
- Strengths of tertiary kinships (**80**).
- Intrablock trends in core radii of first-row elements (Table 11).
- Number of maximum oxidation numbers of first-row elements that are equal to or less than those immediately to their left within their blocks (**11**).
- Number of maximum oxidation numbers of first-row elements that are less than their Groups' maximum oxidation numbers (**11**).
- Percentage of a blocks first-row elements whose subshells of the blocks' predominant type of differentiating electrons are not chemically oxidizable (**91**).
- ΔZ-values of blocks' first triads (**25**).
- First-stage ionization energies of dyads' first elements (**59**).
- First-stage ionization energies of first elements of dyads' second rows (**59**).
- Ratios of heats of vaporization of the last two elements of blocks' first-rows (**91**).
- Intrablock trends in normal boiling points of first-row elements (Table 11).
- Intrablock trends in numbers of hypovalent elements (**91**).
- Percentage of Groups that have common names (**91**).
- Dates of Groups' recognition (**47**).

Appendix XXI
A Dialogue Concerning Two Tabular Expressions
of the Periodic Law

f   enthusiast for the **fdps** table

s   user of the **sfdp** table, He above Ne

Professor William Jolly of the University of California, Berkeley, suggested several years ago that all of the ideas in *New Ideas* be presented in the style of the dialogue below, offered here as a summary and review of this report — and/or an introduction to the Left-Step Periodic Table.

I

f   Why is helium, an s^2 element, grouped with neon, a p^6 element?

s   Helium and neon are both inert gases.

f   Then why aren't nitrogen and oxygen grouped together? Both are diatomic, compound-forming, colorless and odorless, nontoxic, abundant, relatively water-insoluble, low boiling gases.

s   Nitrogen and oxygen have different highest states of oxidation.

f   Oxygen and sulfur have different highest states of oxidation, yet they are in the same group.

s   Oxygen and sulfur have the same number and the same type of valence shell electrons.

f   Mendeleev didn't know that when he placed them in the same group.

s   Oxygen and sulfur have the same lowest state of oxidation.

f   Gold and fluorine have the same lowest state of oxidation, yet they aren't in the same group.

s   When the elements are listed according to their atomic numbers with members of the halogens, alkali metals, alkaline earth metals, and coinage metals lined up vertically, fluorine and gold don't end up in the same group.

f   But helium and neon do, necessarily, end up in the same group?

s   No. But, as mentioned, it makes chemical sense to group them together.

f   That assignment satisfies the First-Element Rule?

s   Which states?

f   First-row elements are less like second-row elements than second-row elements are like third-row elements.

s   In what respect?

f   Atomic properties. With respect, e.g., to first stage ionization energies, electronegativities, proton affinities, refractivitiers, and boiling points, helium is less like neon than neon is like argon?

s   No.

II

f Why is the *s*-block usually placed on the left?

s Created is a horizontal trend, left-to-right, from metals, agents of reduction and formers of basic oxides, to nonmetals, agents of oxidation and formers of acidic oxides.

f Except for the noble gases?

s They're an exception.

f And zinc and copper? Zinc is on the right, nonmetal side of copper, yet *zinc is a better reducing agent than copper.*

s As mentioned, there's another trend, left-to-right: from basic oxides to acidic oxides.

f And the highest oxides of chromium and zinc? Chromium lies on the left, basic-oxide side of zinc, yet *CrO_3 is more acidic than ZnO.*

s The left-to-right rules work if one omits the *d*-block.

f That's part of four periods, out of seven.

s Generally that part of periodic tables is not much discussed in beginning courses in chemistry.

f Why not?

s The most abundant elements in the lithosphere, biosphere, and atmosphere, are, for the most part, among Mendeleev's "typical elements", i.e., the elements in the three early periods of the Conventional Periodic Table, before one gets to the *d*-block.

f The *Periodic* Classification of the Elements is based on terrestrial abundances of the elements?

s Well, then: Why is the *s*-block usually on the left-hand-side of periodic tables?

f For one thing, with that location for the *s*-block, periods start with a group of elements whose maximum oxidation numbers are +1.

s That seems natural.

f Indeed, during the half-a-century or so that the rare gases were thought to be noble, they appeared on the left-hand-side of periodic tables, as Group 0 — despite the fact that numbering, or enumeration, generally does not begin with "zero". There is no "0" among the Roman numerals.

s You're suggesting that by that inconsistency chemists should have suspected a Group oxidation state for the Noble Gas Group of VIII?

f Hindsight's vision is 20/20. Yet, sometimes there's more wisdom in conventional wisdom than at first realized.

s For instance?

f The valence-stroke of classical bond diagrams turned out to represent two electrons. The rule that valence-strokes never cross each other turned out to represent operation of a Principle of Spatial Exclusion for electron-pairs — encoded, if you like, by assigning to electrons a "classically nondescribable two-valuedness", called "spin". It is a bit of a stretch, of course, to suggest that the name hel*ium* implies that its proper location in periodic tables is above a column of metals.

s   And, as you were about to say: For another thing?

f   Placement of the *s*-block on the left-hand-side of periodic tables expresses graphically one of the leading features of Chemical Periodicity: a dramatic change in elements' properties on exiting the *p*-block — with increasing atomic number — and entering the *s*-block.

s   Why, then, would one ever wish to place the *s*-block on the right-hand-side of a periodic table?

f   To give graphic expression to regularities in properties of the elements not expressed by the conventional block arrangement.

s   Such as?

f   For starters: block-to-block trends, particularly in distinctiveness of Groups' first elements with respect to their congeners, greatest in the *s*-block, least in the *f*-block, and in a block-to-block trend in similarities to each other of elements of blocks' first rows, least in the *s*-block, greatest in the f-block.

s   How can the *s*-block be at two places at the same time?

f   By employing a leading construct of human thought: *latency*, called in psychology "potentiality", in quantum physics "The Superposition Principle", in chemistry "Resonance". Latent in the Order of Orbital Occupancy in Bohr's Aufbau Process are different punctuations of the Order, realized, in different situations, as different periodic tables. Below are two of the punctuations.

The Conventional, Long Period = sfdp Punctuation

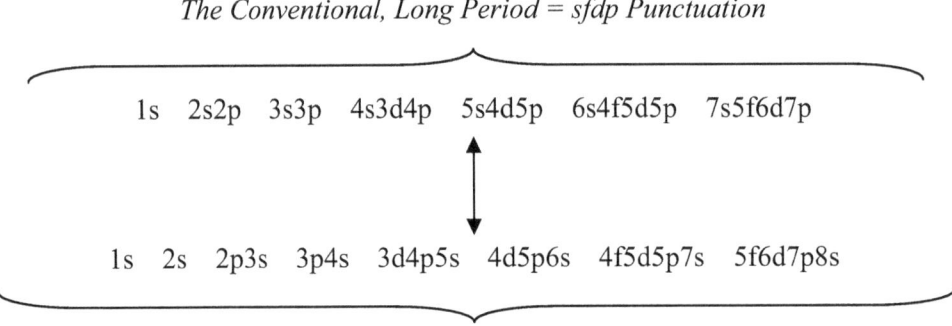

The Left-Step, Long Period = fdps Punctuation

Illustrated is Hegel's dialectic: the art of arriving at the "truth" by disclosing deficiencies in opposing views ("thesis" and "antithesis") and overcoming them (in a "synthesis"). The thesis "s-block on the left" is deficient in that it does not give graphic expression to many of Periodicity's regularities. The antithesis "s-block on the right" is deficient in that it does not give graphic expression to occurrence of major changes in elements' properties on exiting the *p*-block for the *s*-block. The synthesis says: use whichever periodic table is best suited to the situation at hand.

s   Even one that has helium above neon?

f   Perhaps that's the exception that proves the rule?

Appendix XXII

Formation of the ACS Presidential Ad Hoc Committee on Column Labels

The study of periodic tables reported on herein began on receipt of the following letter.

 American Chemical Society

OFFICE OF THE
PRESIDENT

George C. Pimentel
. President-Elect, 1985
President, 1986
Immediate Past President, 1987

Professor of Chemistry
Department of Chemistry
University of California, Berkeley
Berkeley, California 94720
(415) 642-6330

Professor Henry A. Bent
Department of Chemistry
North Carolina State University
Raleigh, North Carolina 27695

Dear Henry:

As President of the American Chemical Society, I have received many adverse comments on the proposed changes in the Periodic Table and I have been urged to look into the matter. It is pointed out that these changes will have profound implications for literature and textbook usage and it would be most unfortunate if the ACS were to set off in a direction that might impede chemistry and that might be rejected by the scientific community.

As a result, I have collected and read all of the comments that have appeared in the Chemical and Engineering News (which ran about 3 to 1 against the new proposals) as well as the various articles that have appeared in J. Chem. Ed., Science Teacher, and C and EN. This groundswell of objection caused me to consider the idea of appointing a Presidential Ad Hoc Committee to review the Periodic Table issue. With that in mind I wrote to a number of ACS Journal Editors and prominent inorganic chemists asking their views about 1) the desirability of appointing such an ad hoc committee, 2) if appointed, what should be the committee's charge, and 3) what should be the committee composition. The answers were unanimous in favor of such a review and several names, including yours, were suggested as desired participants.

So I am writing to ask if you would be willing to serve on this ad hoc committee. Furthermore, I should be especially gratified if you would consent to chair this group. Your research activities in the inorganic area and your national leadership role in chemical education qualify you perfectly to lead this inquiry. The charge to the committee would be as follows:

The Committee should:

1. Evaluate the recommendation of the ACS Committee on Nomenclature as well as alternatives, with attention to the effect of the proposed change on the scientific communities that would be affected: chemistry teachers at all levels, research chemists, teachers and scientists in related fields such as solid state physics and materials sciences, etc.

2. Advise on whether the ACS should proceed with implementation of the recommended changes (in Journals, Chemical Abstracts, etc.) or whether there should be a moratorium on such implementation pending further deliberations and consideration of alternatives.

From Pimentel, 14 Jan. '87

The Players

Kazuo Yamasaki   - Head of IUPAC Committee on Nomenclature of Inorganic Chemistry
Daryl H. Busch   - Chair, IUPAC Commission on Inorganic Nomenclature
Kurt L. Loening   - Chairman, ACS Committee on Nomenclature, Nomenclature Director of Chemical Abstracts
Thomas E. Sloan   - Chairman, Division of Inorganic Chemistry, Nomenclature Committee

The Timetable

A summary of the main events, as derived from articles and letters in C&EN of 1985 and 1986.

1959   IUPAC recognizes problems in format of Periodic Table (PT)

1964   A PT format almost identical to the 1-18 numbering of the 1983 ACS and 1986 IUPAC tables was published by Gunnar Hagg in a book titled "Allman och Oorganisk Kemi", translated by H.T.Evans for Wiley, 1969.

1970   IUPAC recommends a PT format now used rather uniformly in Europe.

1979   ACS Division of Inorganic Chemistry begins study of U.S and European discrepancy in use of A and B for Main and Transition groups.

1981   W.C. Fernelius and W.H. Powell present review of PT formats to ACS Committee on Nomenclature

1982   Report published [J. Chem. Ed. 59(6), 504 (1982)], reviewed at ACS Meeting in Minature (Cleveland, April), and presented by Fernelius at the 7th Biennial Conf. on Chem. Education, Aug. 8-12.

1983   Mar. 24—paper "The Periodic Table: Historical Road to Confusion" given at ACS meeting in Seattle.

Aug. 29—Meeting of Div. of Inorganic Chemistry Committee on Nomenclature boils down all suggestions to 4 formats. After thorough consideration, the one designated "ACS adopted" on summary table is recommended to ACS.

Nov. 4— ACS adopts 1-18 numbering of ACS Committee on Nomenclature.

???? —— Chemical Abstracts announces they will retain U.S. usage through the 11th Collective Index-but will combine both it and 1-18 forms in abstracts, e.q. ".... group 11(IB) metals."

1985   Feb. 4—C&EN article on new format appears. Soon thereafter, letters to the editors began to flow in. See summary of them.

1986   Mar. ——IUPAC approves 1-18 format and seeks comment worldwide.

1987   Sept.——New York state high schools will begin using the 1-18 format.

Appendix XXIII

A Letter to the Editor of *Chemical and Engineering News,*

October 5, 1987

SIR: William B. Jensen (C&EN, Aug. 17, page 2) and many other writers have discussed the practical and logical deficiencies of the proposed ACS-IUPAC table group labels in these pages. The deficiencies are dismaying, but should not be surprising. The decision to use the 1-18 group numbering scheme was a political one, not a logical or practical one.

To explain what I mean by this, let us first suppose that the decision-making authority had been given to a group of organic chemists. They would have approached their deliberations with the precept "carbon is in group four" engraved in their hearts. They would have put some sort of numbering on the transition metals ("Go ahead, put iron, cobalt, and nickel in the same group; you can't tell anything about their chemistry from any conceivable group numbering anyway") and ignored the lanthanides and actinides. The resulting group numbers would be very practical and useful.

Next, let's suppose instead that chemists devoted to the lanthanides and actinides had been appointed. In all likelihood, they would have produced a table with 1-32 numbering. They would proclaim its eminently logical features, say things like, "This change has been long overdue," and shrug their shoulders when confronted with the practical uselessness of a table that puts titanium in group 18 and carbon in group14 (oops, I mean 28).

Instead, the decision was put in the hands of a group of inorganic chemists, who gave their beloved transition metals main-group numbering. It is neither very logical nor very practical [for beginning students of chemistry, and their teachers], but was politically almost inevitable.

As scientists we are devoted to our own systematic ways of pursuing truth. We are often dismayed when politics seems to intrude into our domain, as in the scientific creationism case, and in the removal of certain funding decisions from peer review to the pork-barrel realm. But politics—that is, human relations—is a necessary and inevitable part of life. I hope the ACS will have the wisdom and courage to recognize this, reverse the original bad decision (as the Louisiana legislature, for instance, could not), and turn the search for a better periodic table over to a more broadly based group of chemists.

John P. Sevenair
Professor of Chemistry
Xavier University, Louisiana

Sometimes issues must incubate before their resolution becomes obvious: that, in this instance, latent in Mendeleev's Line are many periodic tables; that latent in a periodic table are many column-labeling schemes; and that latent in each column-labeling scheme are various virtues and drawbacks, *depending on intended use(s).* Recommendations of the chair of the ACS Column-Labeling Committee are, consequently, after these many years, partly political, partly practical, partly logical, and partly scientific. Passage of time suggests, as only it could do, that the periodic-table-using community should *continue with the status quo,* namely: use of the option of no action by the ACS; use of 1-18 labels by professional chemists; and use of one or more of several alphanumeric column-labeling schemes by chemical educators. The scheme favored throughout most of this report is the Nℓ scheme. It is practical and logical. Of course, the same thing could be said of any existing column-labeling scheme, *in particular circumstances.*

Synopsis

Work on this report began with an effort to resolve on scientific grounds an international controversy regarding periodic tables' column labels (Appendices XXII and XXIII). That uproar, it's concluded, was a tempest in a teapot. Having different labels for different purposes — teaching, e.g.; or abstracting — is an asset, as is existence of different periodic tables for different purposes: for display of, e.g., secondary kinships, the *s*-block's singularities, block-to-block trends, or Periodicity's arithmetical regularities.

In creating a periodic table from the Periodic Law, Mendeleev's statement that the Periodic Law is about *atoms*, the conclusion that "Chemical periodicity and the periodic table now find their natural interpretation in the detailed *electronic structure* of the atom" (1), and *regularities* in the electronic structure of the atom, suggest, jointly, use of a table construction convention of *maximum regularity*. Produced is a table of no irregularities: the Left-Step Periodic Table. Its equation of form in terms of its periods' ordinal numbers, P, its blocks' ordinal numbers, ℓ, and its blocks' rows' ordinal numbers, *r*, is -

$$P = r + 2\ell$$

P-values for blocks' first rows ($r = 1$) for $\ell = 0$, 1, 2, and 3 are, respectively, 1, 3, 5 and 7. Described is a step table of step height 2. Needed in order to make it a left- and not, perhaps, a right-step table is the supplementary row-occupancy convention -

for given P, largest ℓ first

The information in the algebraic and verbal statements above can be re-expressed by an analytical expression for the ordinal numbers in periodic tables of blocks' rows, R, in terms of P and ℓ:

$$R(P, \ell) = (P^2 + 2P + \delta)/4 - \ell$$

$$\delta = 0 \text{ or } 1 \text{ as P is even or odd}$$

R-values for different values of P and ℓ are displayed in Figure 56 (following page).

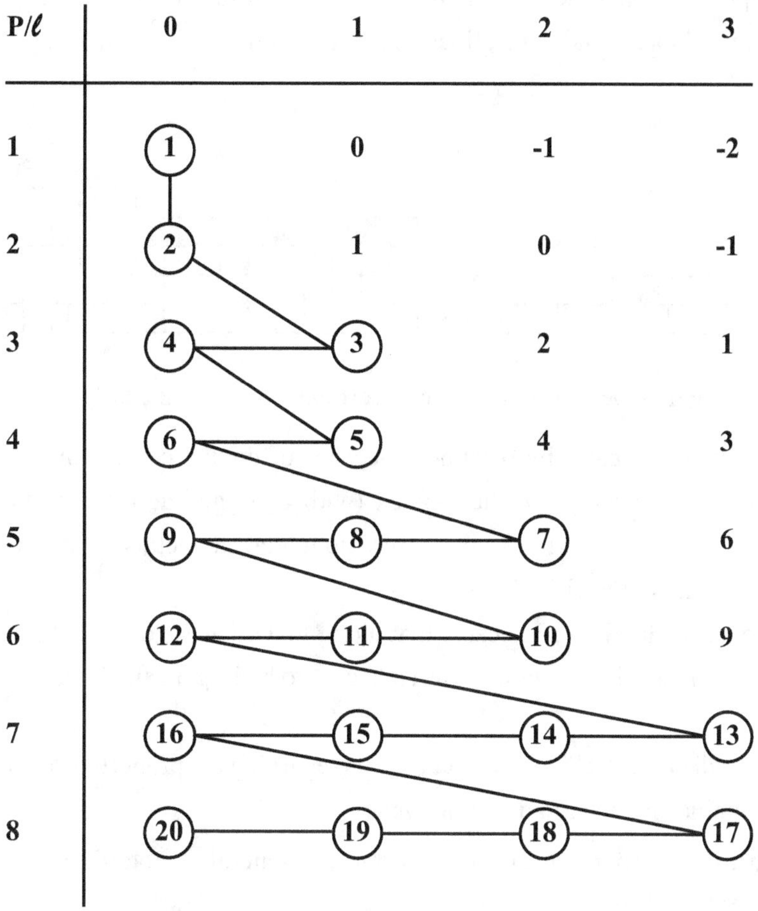

P/ℓ	0	1	2	3
1	(1)	0	-1	-2
2	(2)	1	0	-1
3	(4)	(3)	2	1
4	(6)	(5)	4	3
5	(9)	(8)	(7)	6
6	(12)	(11)	(10)	9
7	(16)	(15)	(14)	(13)
8	(20)	(19)	(18)	(17)

Figure 56. A P/ℓ Spread Sheet for Display of Blocks' Rows' Ordinal Numbers in Periodic Tables.

The string-figure created by connecting first-occurrences of R-values as one reads downward in Figure 56, *right-to-left,* is the mirror image of the second string-figure in Figure 54, read downward, *left-to-right.*

The leading use of the Left-Step Periodic Table is, at the outset, location of the "problem elements" (H and He, La and Ac, and Mg) — with all the consequences pertaining thereto of having helium above beryllium in a periodic table.

One important consequence of He/Be is concordance with a Correspondence Principle between Periodicity's ordinal numbers r and $ℓ$ and nodal properties of orbitals occupied by a block's predominate type of differentiating electrons.

r = number of non-nucleus-containing radial nodal surfaces

$ℓ$ = number of nucleus-containing angular nodal surfaces

= power of a nodal point at the nucleus

Another important consequence of helium-above-beryllium is existence, accordingly, of inter-block trends in intra-block trends in distinctiveness of Groups' first elements.

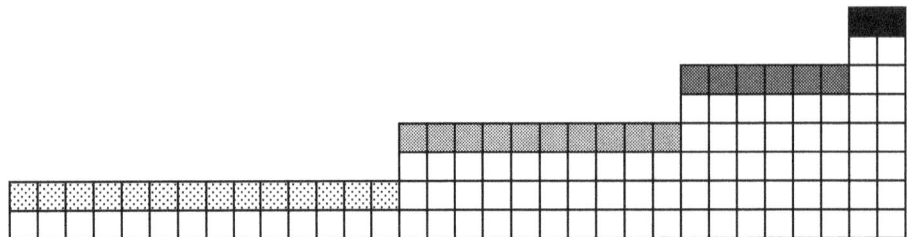

The figure above may be read in two ways: vertically and horizontally.

Read vertically, it indicates by blackness degrees of distinctiveness of Groups' first elements relative to their congeners: high in the *s*-block (with hydrogen above lithium and helium above beryllium), lower in the *p*-block (in its trends from nonmetals to metals), and lower yet in the *d*-block (all elements of which are metals).

Read horizontally, the figure indicates that along blocks' first rows, distinctiveness, element to element, is highest in the *s*-block, with helium adjacent to hydrogen, and lowest in the *f*-block, among the rare earths.

After the Periodic Law itself, leading statements regarding the properties of the chemical elements concern their metallic character and distinctiveness.

- Downward in Groups in all periodic tables, generally metallic character increases and distinctiveness decreases.

- Along sfdp periods of Conventional Periodic Tables, metallic character tends to decrease, left to right.

- Along blocks' top rows, distinctiveness in the Left-Step, fdps Periodic Table increases, left to right.

The fdps and sfdp periodic tables are complementary. The sfdp tables highlight vertical and horizontal trends in elements' metallic character. The fdps table highlights vertical and horizontal trends in elements' distinctiveness.

Notes

(a) Central features of this report, including most of its figures, were presented at the Edelstein Award Symposium honoring William B. Jensen at the 230th ACS National Meeting in Washington, D.C., August 30, 2005, under the full title: "Chemical Periodicity and The Helium Question: To Be or Not to Be? How the Speaker Came to Conclude, Unexpectedly, Following in the Footsteps of Mendeleev and William Jensen, that Chemical Evidence Strongly Supports the Notion that Helium's Natural Position in Periodic Tables is above Beryllium, Not Neon." Titles of earlier versions of the report include "He/Be", "Regularity and Periodic Tables", "f d p s", "Understanding Chemical Periodicity", and "Misconceptions".

(b) Arguably, use of the Convention of Maximum Regularity, chosen to produce the Left-Step Table, widely considered by chemists to be merely an electron-orbital filling chart, not a true periodic table, implies that the so-called "Chemical Capture of the Left-Step Periodic Table" in Section **1** is, in fact, on the chemical side, a hybrid: a *Physical* Chemical Capture". The situation is analogous to Bohr's empirical, chemically-and-physically-guided introduction of the Aufbau Procedure (69). A careful historical analysis of the logic behind the evolution of any route to the Left-Step Periodic Table would probably render a judgment as to whether or not it's purely a chemical or purely a physical route as impossible as unnecessary to make. Actual routes carry both route markers: physics and chemistry. Physical chemistry, Wendell Latimer used to say, is as a method for reaching results, "no holds barred". What's significant is the result: here, a table (or "chart") with overlooked uses in chemistry.

(c) Understanding the significance for chemical thought of He/Be requires pursuit of the implications of the LSPT, which requires He/Be. Similarly, recognition of Periodicity's block-to-block trends requires the mind-set set by the LSPT, justification for use of which resides, in major part, in occurrence of block-to-block trends. It's the old question of the chicken and the egg. Which came first? Recognition of block-to-block trends or recognitions of the chemical significance the LSPT and He/Be? Actually, of course, the three aspects of the Periodic System evolved together. That state of affairs has led to a report that, consequently, begins with an abstract and a longer abstract, synopses and summaries, and that contains numerous numbered sections and references to notes, an annotated bibliography, appendices, figures, legends to figures, and tables. The situation is similar to one that is said to have lead Maxwell and Bohr in their public lectures to attempts to instruct about the correlation of too many things at once (69, p11). In chemistry, few things correlate more things in more ways than the Periodic System, whose scope this report attempts to enlarge, through remarks regarding newly recognized correlations among previously uncorrelated facts.

(d) Zinc, e.g., is a better, not poorer, reducing agent than copper, whereas zinc's highest oxide is less, not more acidic than chromium's highest oxide.

(e) "I just had a nasty thought regarding the phrase 'The Periodic Table of the Elements'," wrote Professor Jensen (private communication, July 2004). "Not only is there a problem with the article 'the' as you rightly point out [some 700 tables exist], there is an even a bigger problem with the word 'elements' . . . The problem is that the word element is no longer synonymous [as it was in Mendeleev's day] with the word atom. . . In short, 'The Periodic Table of the Elements' should really be 'A Periodic Table of Atoms'".

(f) Mendeleev devotes a long footnote in *Principles* (Vol. II, pp22-23) to the principle that the Periodic Law should not be represented by *continuous curves*. Each "period can only contain a definite number of members. For this reason there can be no other elements between magnesium, which gives MgX_2, and aluminum, which forms AlX_3; there is a break in the continuity, according to the law of multiple proportions." The essence of the matter is expressed by periodic *tables* that have *no gaps <u>within</u> their periods*, between, e.g., magnesium and aluminum. It was precisely for that reason that Pauling's periodic table of choice was one similar to that shown in Figure 37. Below are the column labels Pauling placed, at different times, over a period of fifteen years, immediately above his table's first long period, listed at the bottom.

1945

| 0 | Ia | IIa | IIIa | IVa | Va | VIa | VIIa | VIII | Ib | IIb | IIIb | IVb | Vb | VIb | VIIb |

1955

| 0 | I | II | III | IVa | Va | VIa | VIIa | VIII | Ib | IIb | IIIb | IV | V | VI | VII | 0 |

1960

| 0 | I(a) | II(a) | IIIa | IVa | Va | VIa | VIIa | VIII | Ib | IIb | IIIb | IV(b) | V(b) | VI(b) | VII(b) | 0 |

| Ar | K | Ca | Sc | Ti | V | Cr | Mn | Fe | Co | Ni | Cu | Zn | Ga | Si | P | S | Cl | Kr |

Pauling's 1945 suffixes followed the European AB scheme (Figure 41). In 1955 he dropped suffixes in labels of columns that contain conventional "main group elements", except in the case of Groups IIIa and IIIb, where he didn't drop the "b" but did drop the "a". Pauling evidently considered the Al/Sc kinships to be stronger *and* closer than the Al/Ga kinship. Suffixes in 1960 are a compromise between those of 1945 and 1955 (except in the case of IIIa). In 1945, and again in 1960, Pauling connected each element in the Na-to-Cl period (immediately above his Ar-to-Kr period) by tie-lines of equivalent character to two other elements. In 1955 only Al was so connected. Always hydrogen was connected in an equivalent manner to *both* lithium and fluorine. Always La and Ac were located beneath Sc and Y. And always helium was located above neon. The final periodic table presented by the 20th century's leading chemist and chemical educator might be faulted, from the standpoint of the present report, on some thirteen counts. Illustrated is a remarkable phenomenon: a poor understanding by most chemists and chemistry teachers of a central feature of chemical thought. The author himself was one such individual two decades ago. One wonders: What's the reason for virtually universal uncritical use of periodic tables? Early introductions to them, before one is able to think about them critically? And, hence, emotional, not rational, attachment to such questionable features as helium-above-neon? Called to mind is the national scandal regarding college flunk rates in calculus, despite — or because?! — of its earlier and earlier introduction into school curricula. The author recalls, similarly, how impossible it could be to get college chemistry majors to use the definition — hence always valid — H = E + PV after introduction in high school chemistry to the sometimes valid relation $\Delta H = Q_p$. Early exposure to mankind's great thoughts may be a dangerous thing. "Drink deep, or taste not the Pierian spring," advised Pope. "There shallow draughts intoxicate the brain . . ."

(g) After an introduction to the entropy-function as nothing more, nor less, than an empirical, numerical measure of irreversibility, a p. chem. class was asked: "Any questions?" Silence. Then, a response: "How can you argue with the Second Law?" One might suggest, however, that it might be nice to have an explanation. Of course, "[J]ust as without knowing the cause of gravity it is possible to use the law of gravity," noted Mendeleev, in a continuation of famous footnote 11 of *Principles* (Volume II, p24), "so for the aims of chemistry it is possible to take advantage of the laws discovered by chemistry without being able to explain their causes." What's possible, however, may not be fully satisfying. A leading goal of this report is to have atomic orbitals' circle-diameter diagrams (Figure 29), the Row-Orbital Correspondence (**48**), and the physical explanation of late occupancy of high-ℓ orbitals (**61 – 64**) be seen to stand in an explanatory mode to the empiricism of chemical capture of the LSPT (Figure 1) and Periodicity's integers (Figure 21) and the Row-Occupancy Rule (**5**) as, e.g., Boltzmann's relation $S = k\ln W$ stands to the Second Law of Thermodynamics.

(h) IUPAC's abolition with its 1-18 column labeling scheme of any indication via column labels of secondary chemical kinships (unless one considers the digit "1" of column labels in the teens in the p-block to be part of a prefix-numeric column-labeling scheme) has had the consequence, points out Professor Jensen (private communication), that secondary kinships, although widely discussed at the outset of the history of the Periodic Law by Mendeleev in "Principles", are being rediscovered by younger generations of chemists and published as new insights into Chemical Periodicity.

(j) Landis and coworkers "have recently developed a far-reaching extension of localized hybridization and Lewis-like bonding concepts" that reveals "many striking parallels" between familiar sp-hybridization in the p-block, where most secondary kinships begin, and sd-hybridization in the d-block (45), where they end. Extrapolation to include the f-block yields for the number of s + p, d, f hybrids involved 1 + 3, 5, 7. Numbers of main lobes per sp, sd, and sf hybrid are, respectively, 1, 2, and 3. Numbers of valence-shell "domains" are, accordingly, $1(1 + 3) = 4$, $2(1 + 5) = 12$, and $3(1 + 7) = 24$. [The general case is $\ell\{1 + (2\ell + 1\}$.] Corresponding coordination polyhedra of electron "domains" are in the sp case the well-known tetrahedron and in the sd case an icosahedron, which yields bond angles surprisingly close to those calculated (45) for, e.g., WH_6 (46).

(k) The same method is used with the periodic table that appears on the cover of volume I of the new (second, 2006) edition of Wiley's "Encyclopedia of Inorganic Chemistry", edited by R. Bruce King.

ADDED NOTE: In a gem of clear reasoning and scholarly insight, professor William Jensen, in an article titled "The Positions of Lanthanum (Actinium) and Lutetium (Lawrencium) in the Periodic Table," *J. Chem. Ed.*, **59**, 634-636 (1982), terminated, one might have thought, the "long, but poorly publicized, history" of the debate "as to whether lanthanum and actinium or lutetium (and, more recently, lawrencium) should be placed in the [scandium-yttrium] group." A major goal of this report is to accomplish for He what professor Jensen accomplished for Lu and Lr. Below are his article's concluding remarks.

> "[I]n talking with his fellow chemists, the author discovered that none of them was aware of the [physical] evidence favoring the reassignment of lutetium and lawrencium or indeed that there was any question about their placements (a category in which the author must include himself until very recently). *As chemists, the periodic table is presumed to be our special province; <u>surely it's about time we pay attention to what the physicists have to tell us about its arrangement</u>*" (emphasis added).

Physics tells us that helium is not related to neon by a primary kinship and that it is related to beryllium by some kind of kinship. Both Jensen and this report consider the He-Ne and H-F relations to be tertiary ("isoacceptor") kinships. Jensen's definition of "isovalent" (primary) kinships makes the He-Be and H-Li relations "isodonor" (secondary, not primary) kinships. His definitions fit nicely a step-pyramid table that has the s-elements on the left and, consequently, H and He in an "H-He Block" of their own, as a monad above three dyads. That table was brought to the present author's attention through professor Jensen's studies, prior to knowledge of his current definitions of chemical kinships. Similarly, the LSPT came to the author's attention before the idea (following in professor Jensen's footsteps) of viewing helium's kinship with neon to be a non-primary, tertiary kinship. Suggested is the thought that, in general, in the case of the highly distinctive, light elements, a periodic table preceded their group assignments. That was not true, however, for B, C, N, O, and F. Some chemists had grouped those elements with Al, Si, P, S, and Cl, respectively, prior to 1869, the date of Mendeleev's first periodic tables. Not included in that list of light elements, however, is Nature's lightest element, hydrogen. A cursory survey of the literature suggests an answer to -

A QUESTION FOR HISTORIANS OF CHEMISTRY: Which came first for hydrogen: Its conventional assignment, in the alkali metals group, or a periodic table?

De Chancourtois' "telluric helix" of 1862 does not have H above Li, Na, and K. Rather, H appears (with slight juggling) above F and Cl. Likewise, Newlands' "octave arrangement" of 1865 aligns H with F and Cl. Odling's "triplet groups" of 1868 do not associate H with either the halogens or the alkali metals, nor does Hinrich's classification of 1869. Meyer's system of 1865 omits hydrogen. (In other respects it's identical to the arrangement of elements in the p- and s-blocks of the Left-Step Periodic Table.)

Of course, it's difficult to prove a negative: that hydrogen was not grouped with the alkali metals prior to Mendeleev's periodic tables of 1870. Tentatively, however, that appears to be the case. Maximum states of oxidation, alone, were insufficient to place hydrogen in the alkali metals group, in the absence of the Periodic Law. As in the case of La (and Ac) and Lu (and Lr), in placement of Nature's two lightest elements in periodic tables, it helps to "pay attention to what the physicists have to tell us about its arrangement."

(m) Despite an antipathy regarding gaps within periods, Mendeleev accepted, at times, a hydrogen-helium stretch — the only one in his entire system. His table "The Periods of the Chemical Elements" begins with the arrangement of atoms pictured in Section **100** rotated clockwise ninety degrees, then flipped, right-to-left. On the other hand, in his accompanying "short-form" table in Volume I of *Principles*, titled "The Periodic System of the Elements in Groups and Series" (pxviii), Mendeleev indicates explicitly by dashes supposed existence of six undiscovered elements between hydrogen and helium, for presumably two reasons: (i) avoidance of gaps in his first "series"; and (ii) completion of his arrangement's overall chess board character.

(n) We aren't born knowing that sensory impressions produced by our surroundings correspond to what we later learn to call *things*. Chemists initially considered Davy's claim regarding the elemental character of that thing that he called "chlorine" to be merely his personal theory of the pale green gas obtained from sea salt. Today we take its elemental character for granted. Physicists at first didn't believe Joule's production of heat from work. Today they define the "calorie" mechanically. Facts and definitions are where theories and hypotheses end up when they are so successful and so familiar for being so useful that we don't give them a second thought. Factness resides in the mind of the beholder. An idea that for one person is exceedingly useful (hence, a fact) may be for another person (in a different line of work) not useful at all (hence, a theory).

(o) Present in chemistry teaching that follows the precepts of Lavoisier (72) and Whitehead (73) and that, accordingly, features lecture demonstration-experiments (74) are many opportunities to refer to kinetic-molecular theory, the first and second laws of thermodynamics, proton-transfers, more complex reaction mechanisms, bond diagrams, and periodic tables. Presented, also, but largely overlooked by chemical educators, and by the present author during most of his years in academia, are excellent opportunities to provide observers with what William Zinsser has called "break through" exercises in expository writing (75): "Describe how something works." All I want, Zinsser says, are simple, short, declarative sentences that follow each other logically. In the observed demonstration-experiment, what was used, done, seen, said, written, drawn, imagined, concluded, and tried next?

(p) The author's view of Periodicity from professor Jensen's shoulders differs, of course, in some details, from his view of it. From the same evidence the stander and standee have reached complementary conclusions regarding, e.g.: definitions of chemical kinships; the character of the kinship between helium and beryllium and between magnesium and zinc; use of the word "block" and the phrase "main group"; theoretical significance of dyads and triads; and whether or not Janet's arrangement of elements' symbols is a ("true") periodic table or is more accurately described as an "electron configuration filling chart". On the most important issue for the present report, however, we stand in agreement: *Helium is not a congener of the Noble Gases.* "While it is true that some chemists incorrectly argue for placements (e.g. H in the halogen group, He in the noble gas group) based on this criterion ['that placement of the elements in a periodic table is based on the physical properties of simple substances']," wrote professor Jensen February 20, 2006, "this is an error on their part and reflects their lack of understanding concerning the basis of the classification in the table (an unhappily frequent occurrence in the literature dealing with the table)."

Annotated Bibliography

1. Greenwood, N. N. and Earnshaw, A. *Chemistry of the Elements;* Pergamon Press, New York, 1985, Chapter 2, p24. This highly regarded textbook is an excellent account of its titular topic. Displayed in Fig. 2.5 are elements formal oxidation states "in a format originally devised by Mendeleev in 1889". It is the basis of the first part of this report's Table 1 and a chemical route to the Left-Step Periodic Table (**11**). Cited as the first of ten trends to be discussed is the trend central to this report (**30**): "The 'anomalous' properties of elements in the first short period (from lithium to fluorine)". The authors' second and third cited trends, "anomalies" associated with "d-block contraction" and "effects of the lanthanide contraction", are consequences of Periodicity's dyadic character, the most prominent feature of the leading topic of the present report: the Left-Step Periodic Table, particularly its location of helium above beryllium.

2. Anonymous reviewer for *J. Chem. Ed.* of an early version of a portion of the present report. The helium-neon electron configuration inconsistency in an (otherwise) up-to-date version of the Conventional Periodic Table is unique. Nowhere else in the table does such an inconsistency occur. It might appear, at first glance, to be merely a minor, inconsequential glitch, and it would be, were it not for the fact that, as part of a periodic table, helium is part of a *system of connections*, "The Periodic *System,*" in which every part, in its proper location, is connected in numerous ways to other parts. Helium's improper placement above neon creates in the System twin blemishes: a vacancy were helium should be; and a wart where it should not be. Both blemishes render invalid most of the generalizations regarding the Periodic System cited in this report.

3. Mazurs, E. G. *Graphic Representation of the Periodic System During One Hundred Years;* University of Alabama Press, University, Alabama 35486, 1947, p93. Mazurs book is the leading compendium of periodic tables. It contains many references and a thorough "Introduction to Electron Configuration Tables" and their "Basis in Atomic Structure". "The time has come," asserts Mazurs, p136, "to accept electron configuration tables" as one of several useful periodic tables. Mazurs answer to the question "Which one [of "a multitude of periodic tables"] is the best table?" is the same as the present author's answer (**99**): "[It] depends on the purpose for which the tables are to be used." Mazurs identifies three uses: exhibition of chemical valence, electron configurations, and periods' shell and subshell structure. The LSPT, considered by Mazurs to be "an electron configuration table", can exhibit "chemical valence" via column labels. In 1899 Mendeleev had remarked that "One can infinitely vary the appearance or shape of the classification of the elements based on the periodic law" (22, p225). Existence of so many "layouts" reflects the fact that the periodic tables stand to Chemical Periodicity rather the way Euclidean space stands to hyperbolic space. Hyperbolic space can be only be partially represent in Euclidean space, but in many ways (25). Chemical Periodicity can be only partially represented by a periodic table, but in many complementary ways.

4. Emsley, J. *Nature's Building Blocks;* Oxford University Press, New York, 2001; p526. The classical "problem elements" bracket the Left-Step Periodic Table: La and Ac at the lower left, H and He at the upper right. To that short list should be added zinc (**90**). Historically, all the heavier elements involved in strong secondary kinships (**79**) were problem elements, as were elements with common lower states of oxidation, particularly Ag(I), Tl(I), Pb(II), and Cr(III). Even to this day universal agreement does not exist regarding the role in the Periodic System of the entire lanthanide and actinide "series" (**92**). Locations in the System of about a third of the known elements are not entirely secure, in the conventional view.

5. Mendeleev, D. "The Periodic Law of the Chemical Elements", *The Chemical News*, Vol. XL, Dec. 26, 1879, p303, reprinted in *Classical Scientific Papers: Chemistry, Second Series: Papers on the Nature and Arrangement of the Elements;* Knight, D. M., editor; American Elsevier Publishing Co., Inc., New York, 1970, p285.

6. White, H. E. *Introduction to Atomic Spectra;* McGraw-Hill Book Company, New York, 1934, pp301-2. Chapter VII contains a discussion of "penetrating and nonpenetrating orbitals" highly relevant to the physical explanation offered in Section **61** for the shapes of periodic tables.

7. Mazurs (3), pp139-140. The Mendeleevian Line at the top of Figure 1 is essentially isomorphic with well-known plots of atomic volumes and first-stage ionization energies of the elements (1), pp28-29.

8. Whewell, W. *The Philosophy of the Inductive Sciences;* Frank Case and Company, London, 2nd edition, 1847, reprinted 1967, in two volumes, Vol. II, p539. Whewell illustrates his "Maxim by which all Systems professing to be natural must be tested" with an example from the vegetative kingdom. The same classification was obtained whether one considered reproductive or nutritive organs. He wonders (p540) "whether this Idea of Natural Affinity is applicable to inorganic as well as organic bodies" and concludes that it may apply to minerals. The time was not yet ripe, however, for an application of Whewell's "fundamental maxim" — that, to be considered natural, *"arrangements obtained from different sets of characters must coincide"* — to a classification of the chemical elements based on different characteristics, such as atomic weights and maximum states of oxidation (Mendeleev), molar volumes (Meyer), first-stage ionization energies, and electron configurations, since not even atomic weights were known with certainty in 1847.

9. Simmons, L. M. "A Modification of the Periodic Table," *J. Chem. Ed.*, **24**, 588-591 (1947). Simmons' "modification" is the LSPT with, however, H above F and He above Ne. Simmons notes a "gradual change from strong horizontal relationships at the left of the chart [sic] [particularly among the rare earths], to strong vertical relationships at the right [as evidenced, e.g., by the prevalence there of named groups]". In a subsequent article, *JCE*, **25**, 658-661 (1948), Simmons displays H and He in the format of the LSPT, which he calls "the Arithmetic Table". Concerning it he says (contrary to the spirit of the present report): *"From the point of view of its chemical properties, there is no justification for grouping helium with [the alkaline earth metals'] column*

. . ." (emphasis added). "The positions of H and He in [the] Table," Simmons continues, "are defended on grounds which are configurational, not chemical."

10. Jensen, W. B. "Classification, Symmetry, and the Periodic Table," *Comp. and Math. with Appls.,* Vol. **12B**, 487-510 (1986). This classic by one of the world's leading authorities on Chemical Periodicity contains the highest density and largest number of significant insights of any account of that subject the present author has seen, after those by Mendeleev himself.

11. Knight (5), p236. Cooke adds (p237): "The difficulty with most of the classifications [of the elements] is, undoubtedly, that they are too one-sided, based upon one [or two] set of properties [such as, in the case of helium and neon, inertness and boiling points] to the exclusion of others . . . This is the difficulty with the old classification into metals and metalloids, which separate [sometimes, in Cooke's day] phosphorus and arsenic, sulfur and selenium, because arsenic and selenium have a metallic luster, while phosphorus and sulfur have not. For a zoologist to separate the ostrich from the class of birds because it cannot fly, would not be more absurd, than it is for a chemist to separate two essentially allied elements [such as beryllium and helium] because one has a metallic luster and the other has not."

12. Knight (5), p260. Newlands devotes most of his short paper on "natural" families or groups to a consideration of triads, particularly one involving magnesium, zinc, and cadmium, a non-tradition triad (**84**).

13. Knight (5), pp285-6. In this short article, Mendeleev discusses two topics: artificial and natural classifications; and "atomanalogies", particularly one involving diagonal relationships: Li : Mg = Be : Al = B : Si. Those are not good analogies from the standpoint of the LSPT. Nor are they supported by first stage ionization energies. Somewhat more "natural" is the following block-to-block trend in ratios of first-stage ionization energies of elements of blocks' first elements to those elements immediately below and to the right of them in periodic tables.

H/Be		B/Si		Sc/Zr		La/Th
1.46	>>	1.02	>	0.96	>	0.80

14. Jensen, W. B. "Where did the term transition metal come from?", *Chem 13 News,* February, 1981.

15. van Spronsen, J. W. *The Periodic System of the Chemical Elements: A History of the First Hundred Years;* Elsevier Publishing Co., New York, 1969. This valuable source of information discusses many interesting ideas, often without, however, reaching definitive conclusions — or ones in agreement with those of the present report. The book features on the insides of its front and back covers a "long form" of the Conventional Periodic Table (usually attributed to Werner) that, from the standpoint of the LSPT, places, correctly, He above Be, but places, incorrectly, Sc and Y above La and Ac. Illustrated is the present author's remark that "Irregularities beget irregularities."

16. Cronyn, M. W. "The Proper Place for Hydrogen in the Periodic Table," *J. Chem. Ed.*, **80**, 947-951 (2003). Cited are a number of interesting similarities between the chemistry of hydrogen and the chemistry of the carbon-silicon Group. From the standpoint, however, of the Rule of First-Element Distinctiveness (**30**), those similarities argue *against* making hydrogen a congener of carbon and silicon. And, on the other hand, differences between the chemistry of hydrogen and the chemistry of the alkali metals, cited as reasons for *not* placing hydrogen at the head of the alkali metals, *are* consistent with the Rule of First-Element Distinctiveness. Overlooked, also, is the fact that the periodic table was first and foremost, at the outset, a table of maximum states of oxidation, regardless of whether or not elements with the same maximum oxidation numbers were metals (e.g. aluminum, lead, bismuth, and tellurium) or nonmetals (respectively: boron, carbon, nitrogen, and oxygen) (**19**).

17. Dobereiner, J. W. "An Attempt to Group Elementary Substances According to their Analogies," Poggendorfs' *Annalen der Physik and Chemie*, **15**, 301-307 (1829). Reprinted in Leicester, H. M. and Klickstein, H.S. "A Source Book in Chemistry," Harvard University Press, Cambridge, MA (1968). Dobereiner did not include in triads Periodicity's first-row elements C, N, O, and F. During the period between the studies of Dobereiner and Mendeleev many chemists devoted much attention to triads (15, pp69-95).

18. Mazurs (3, p72) and van Spronsen (15, p150).

19. Mendeleev, D. *The Principles of Chemistry,* 3rd English Edition, translated from the 7th Russian Edition by G. Kamensky, in two volumes, Longman, Green, and Co., New York, 1905, reprinted by Kraus Reprint Co., New York, 1969, Vol. II, p18, footnote 8. Later Mendeleev says (p20): *"The elements having the lowest atomic weights, although they bear the general properties of a group, still show many peculiar and independent properties."* (emphasis added). Mendeleev himself did not apply that principle to helium, although conceivably, in retrospect, he might have, through use of triad rules (**25**), applied to elements' ordinal numbers in his System.

20. Jensen (10), p506. Three principles have strongly influenced the evolution of the present report. They are: Mendeleev's Absolute Distinction (**34**), Mendeleev's Rule of Light-Element Distinctiveness (**30**), and Jensen's extension of Mendeleev's Rule (**31**). From the standpoint of this report, the Mendeleev-Jensen Rule stands out as the broadest generalization in inorganic chemisry regarding the properties of the chemical elements after the Periodic Law itself.

21. Bent, H. A. and Weinhhold, F. "Periodicity Symbols, Tables and Models for Higher-Order Valency and Donor/Acceptor Kinships," prepared for submission for publication.

22. Mendeleev, D. "How I Discovered the Periodic System of Elements," translated and reprinted in "Mendeleev on the Periodic Law: Selected Writings, 1869-1905," Selected and Edited by William B. Jensen, Dover Publications, Inc. New York, 2005, p193. Included in this valuable collection are Mendeleev's famous Faraday Lecture of 1889, his chapter 15 in *Principles of Chemistry* on "The Grouping of the Elements and the Periodic Law", and three perceptive essays by Professor Jensen

concerning Origins of the Periodic Law, Priority Disputes and Confirmations, and Acceptance and Recognition. In his Faraday Lecture, Mendeleev notes that "periodicity was . . . a direct outcome of the stock of generalisations and established facts which had accumulated by the end of the decade 1860-1870 [particularly regarding] the numerical value of atomic weights. . . Ten years earlier such knowledge did not exist. . . I vividly remember the impression produced by [Cannizzaro's] speeches [at Karlsruhe in 1860], which admitted of no compromise. . ." The present report is a direct outcome of the stock of generalizations that have accumulated [since 1889].

23. Mendeleev (19), Vol. II, p19. The oxide sequence embodies the chemical essence of the Periodic Law. The most important thing to be noticed here, says Mendeleev, is that, as the atomic weight increases the series of elements of small period of seven groups from I. = R_2O to VII. = R_2O_7 periodically repeats itself. Another important thing to be noticed is that in each "small period" some elements are metals, some nonmetals. To judge from trends upward in the f-, d-, and p-blocks in horizontal trends in metal-nonmetal character within blocks, one might suppose that at the top of the s-block each element would be nonmetallic, particularly at the block's upper right corner. Indeed, from the standpoint of an inter-block trend in intra-block trends in metal-nonmetal character, the element at that location in periodic tables should be the most nonmetallic element in the Periodic System, in the sense of being the element most difficult to oxidize.

24. Jensen, W. B. "The Periodic Law and Table," Collected Papers, No. 81, Published in *Britannica on Line*, Encyclopedia Britannica, Chicago, IL, 2000. This scholarly discussion contains an account of the "Electrification of the Periodic Law". In 1911, notes Jensen, a Dutch school teacher, van den Broek, gave a physical interpretation to the chemical ordinal number by suggesting that it corresponded to the positive charge on the nucleus of the atom. Displayed is Sanderson's "double-appendix" table, which makes Mg a congener of both Ca and Zn. Critiqued is IUPAC's 1-18 group-labeling scheme: "Group labels now function as enumerators rather than as descriptors, since they no longer provide information about valence electron counts. Unfortunately [for chemistry students and teachers] this method is largely tied to the peculiarities of the medium-block form of the periodic table. It totally ignores the 14 groups in the F-block . . ."

Jensen carefully distinguishes between use of a letter of the alphabet (such as the sixth one) in a description of an electron configuration of an atom and use of the same letter to indicate a block of elements in a periodic table. The notation and its context are mutually redundant, like the period, extra space, and capital letter between two sentences. Emphasized is an important distinction: two uses of the same letter. Lost, somewhat, is explicit expression of the notation's etymology. Illustrated once again is a characteristic feature of the Periodic System: a richness of expressive forms. The periodic table, column and block labels, and kinship definitions one uses may depend on the uses to which those features of the System are put, and one's familiarity with them.

The situation with the symbols F- and *f*- is analogous to that regarding the symbol "C" in graphic formulas of molecules. Initially the alphabet's third letter, capitalized, stood, in bond diagrams, for *an atom of carbon*. In Lewis's identification of the valence-stroke as two electrons, "C" stands for *the core of a carbon atom:* C^{+4}, written by Lewis "**C**". Writing "f-block" in this report instead of "F-block", or "Group IIIf" instead of "Group IIIF", is, thus, like using "C" in bond diagrams instead of "**C**". Familiarity breeds "contempt" for linguistic niceties. A professional golfer, it's said, can drive, chip, and putt quite nicely, thank you, with one club.

The chief reason for favoring in this report "*f*-block" and "III*f*" over "F-block" and "IIIF" is the prohibition: *Do not introduce new symbols unnecessarily*, combined with the frequent occurrence in English (and other languages) of words that have multiple meanings.

25. Penrose, R. The Road to Reality: *A Complete Guide to the Laws of the Universe*, Alfred A. Knopf, New York, 2005, p805. Finally!: An adequate response to the essay questions that used to bedevil undergraduates when the author was a student at Oberlin College: "Describe the universe and give two examples." Penrose cites as "[e]vidence of electron spin" not Pauli's "2" of the leading factor of the expression $2(2\ell + 1)$ (Appendix II), but, rather, the "2" of an electron-pair bond, p623.

Penrose's "very simple coordinate change", from x and y to X = x and Y = x + y, p189, is exactly analogous to the coordinate change in going from Madelung's diagram (x = ℓ, y = n) to the LSPT's diagram [X = ℓ, Y = n + ℓ (= P_{fdps})] (Figure 30).

His remarks, p40, regarding existence of various different models of hyperbolic geometry, expressed in terms of Euclidean space, calls to mind existence of various tabular expressions of Chemical Periodicity, expressed in two dimensions. It "serves to emphasize the fact that these are, indeed, merely 'Euclidean models' of hyperbolic geometry and are not to be taken as telling us what hyperbolic geometry actually *is*. No one of the models is to be taken as the 'correct' picturing of hyperbolic space at the expense of the others" (emphasis added). Likewise, no one of the dozen or so leading periodic tables cited in this report is to be taken as the 'correct' tabular expression of Chemical Periodicity at the expense of the others (**99**).

26. Jensen's "nasty thought" (e) expresses in a nutshell the central thesis of this report. The key to understanding the Periodic Classification of the Elements is keeping in mind that it's a classification of *atoms*, not simple substances.

27. (22, p193). Mendeleev adds a note in the spirit of Chapman's words quoted at the outset of this report: "In the particular case of carbon, chemical nomenclature clearly expresses the difference between carbon as a simple . . . body (charcoal) and carbon as an *atom* (carbon). For many other elements, indeed for the majority, this distinction of terms does not exist. Thus one designates by the name oxygen both a gaseous body and the same element as a component of water and many other liquid and solid bodies. One would hope that this inadequate terminology will be perfected in time." For the time being, when referring to an element in a periodic table, one might preface the element's name by the adjective *atomic*. One might say, e.g.: "*Atomic* helium [an s2 system] belongs in Group II." That sounds better, to outer and inner ears, than "Helium [an inert gas] belongs in Group II."

28. Knight (5), p267. Anticipating Mendeleev, Odling wrote in 1864 that "With ease [a] purely arithmetical seriation [by atomic weights] may be made to accord with arrangement of the elements according to their usually received groupings", p265. Included among his "usual received groupings" are: B Al (U); C Si – Sn; N P As Sb Bi; O S Se Te; F Cl Br I; Mg Ca Sr Ba; K Rb Cs; Ag Au; and Zn Cd Hg — i.e., a majority of the elements highlighted in color in Figure 1.

29. Mendeleev (5, p285). "[T]here exists between K and Na, in spite of the small difference between their atomic weights, more difference of properties than there is between K, Rb, and Cs."

30. Linnett. J. W. *The Electronic Structure of Molecules: A New Approach,* John Wiley and Sons, Inc., 1964.

31. Hakala, R. "The Periodic Law in Mathematical Form," *J. Phys. Chem.,* **56**, 178-181 (1952). Attention is focused on Janet's LSPT. Displayed in that format are atomic orbitals' $n\ell$ symbols and, in a full table, most of the integers of Figure 20 with, in addition, the principal quantum number n indicated, somewhat as in Figure 31.

32. Shchukarev, S. A. "New Views of D. I. Mendeleev's System. I. Periodicity of the Stratigraphy of Atomic Electronic Shells in the System, and the Concept of Kainosymmetry," translated from Zhurnal Obshchei Kimii, Vol. 47, No. 2, February, 1977, pp246-259, Plenum Publishing Corporation, 227 West 17th St., New York, NY 1011. Scholarly and original reflections on what is called in the present report "first-row (or first-element) distinctiveness". "The adjective 'kainos'," says the author, "means 'new' and by the kianosymmetrics we understand elements containing electrons in the [sub]shell of the atom which for the first time (in the order of increasing atomic number) introduce new angular symmetry (new values of the second quantum number ℓ) into Mendeleev's system." The subshells in question are composed of the orbitals 1s, 2p, 3d, and 4f. They correspond to the first-rows of the *s-, p-, d-,* and *f*-blocks. Row ordinal number (and orbital radial quantum number) r equals 1. The 1s, 2p, 3d, and 4f orbitals have no radial nodal surfaces (except for ones at infinity). They have, continues Shchukarev, "zero value of the 'pseudopotential,' i.e., the special circumstances of a wave nature characteristic for a multielectron shell of an atom and acting on the electron under examination like a force repelling it from the nucleus toward the surface of the atom. [Consequently] 1s, 2p, 3d, 4f, and 5g electrons on the occupation of their vacant orbitals are able to penetrate deeply into the region close to the nucleus . . ." That statement agrees with professor Weinhold's views reported in Section **32**.

33. Paneth, F. A. *Chemistry and Beyond,* Interscience Publishers, New York, 1964, p34. Paneth devotes chapter 6 to "a very important epistemological distinction which astonishingly enough has received very little attention in [the] literature", namely: "the epistemological status of the chemical concept of 'element'", especially the double meaning for it that Mendeleev had in mind, designated by Paneth *simple substance* (corresponding to chemistry's qualitative, "naively realistic", descriptive, "Daylight View" of the world, apprehended by our senses) and

basic substance (corresponding to physic's "Night View", taken in the present report to stand for *atoms*). Paneth ends his essay, on the "philosophical principles of our science" (which, he emphasizes, Mendeleev regarded as the main substance of his textbook), with praise for "the undiminished freshness" of *Principles of Chemistry*, "which every chemist may read with great profit even today."

34. Mingos, D. M. P. *Essential Trends in Inorganic Chemistry,* Oxford University Press, New York, 1998. This informative, well-illustrated, up-to-date, and generally highly readable, 392-page textbook remarks, p33, that "The number of elements in successive rows is given by the formula $2[INT\{(N+2)/2\}]^2$ where N is the number of the row (N = 1-7) and the function INT(x) represents the rounded-down integer x." Curiously, although the book features Madelung's Diagram on its cover, it does not display the mathematically equivalent and chemically and physically useful Left-Step Periodic Table. Consequently, it does not contain dyads' ordinal numbers, D, useful in stating lengths of "rows", as simply $2D^2$. Nor (of course) does it cite, among its "essential trends", block-to-block trends (since helium is located above neon in its periodic tables). Included is a major section on secondary, Group N/Group (N+10) comparisons. On the inside of the front cover is a Conventional Periodic Table. Hydrogen is above both lithium and fluorine. Footnoted lanthanides and actinides start with La and Ac and end *fifteen* elements later with Lu and Lr. Absent, as usual, is any indication as to which elements are congeners of Sc and Y: La and Ac or Lu and No? Chemists, one is tempted to conclude, are, surprisingly, often remarkably cavalier in their treatment of periodic tables.

35. Wiswesser, W. J. "The Periodic System and Atomic Structure. I. An Elementary Physical Approach," *J. Chem. Ed.,* **22**, 314-321 (1945). Wiswesser gives the following account, unknown to the present author at the time he was coming to the conclusions cited in Section **61**, of the physical basis for Madelung's parameter $n + \ell$: "All linear nodes [angular nodal surfaces] slice through the nucleus and therefore deny an electron to this strong bonding region. Spherical [nodal surfaces], on the other hand, allow a high probable occurrence in this nuclear region. Consequently, when inner groups of electrons and high nuclear charges are present, these *linear nodes keep the electron almost twice as far from the nucleus as spherical ones.* This difference is especially marked when the electron might penetrate inner layers of electrons and thereby experience a higher nuclear charge which these inner electrons otherwise would neutralize. Except for this difference between linear and spherical nodes, patterns with the least number of nodes (prohibited regions) will be occupied first. Counting linear nodes twice for their relative value, *the pattern* [of energy levels] *will be filled in increasing order of $n + \ell$* (rather than n alone)."

36. Pauling, L. and Wilson, E. B. *Introduction to Quantum Mechanics,* McGraw-Hill Book Company, Inc., New York, 1935, Table 21-3. The authors write the relation among atoms' quantum numbers as $n = n' + \ell + 1$, where *n* is the "principal quantum number" and ℓ is the "azimuthal [angular] quantum number." They call n' the "radial quantum number". It is the number of radial nodal

surfaces *not counting the one at infinity*. That is to say, $n' = r$ (this report's "radial quantum number") – 1.

37. Lowdin, P. O. "Some Comments on the Periodic System of Elements," *International Journal of Quantum Mechanics.* Symposium 3, 331-334 (1969). "[I]t is perhaps remarkable that, in axiomatic quantum theory, the simple energy rule (order of filling of orbitals) has not yet been derived *from first principles*" (emphasis added).

38. Mendeleev (19), Volume II, p329.

39. Jensen (24) p7. Atoms are said to be "isovalent" when they have the same valence electron count (e) and the same valence vacancy count (v) and, hence, the "same overall valence manifold (m)": m = e + v. Jensen's m-values are: for H and He, 2 + 0 = 2; for conventional main group elements, 8; for elements of the d-block, 18; and for elements of the f-block, 32. Hence, hydrogen and helium (m = 2) are not "isovalent" with (and, consequently, are not congeners of), respectively, the alkali and alkaline earth metals (m = 8). In Jensen's scheme, oxygen, e.g., and sulfur have identical valence manifolds, 8, although they have different maximum ligancies.

Electron-domain models of conventional valence-stroke diagrams of OF_2 and SF_6 place the oxygen core O^{+6} of OF_2 in the tetrahedral interstice of four tetrahedrally coordinated "electride ions", or electron-pairs, of two O-F bonds and two oxygen lone pairs and the sulfur core S^{+6} in the octahedral interstice of six octahedrally coordinated "electride ions", or electron pairs, of six S-F bonds.

In those models, the sulfur core S^{+6} (not shown, on the right) has a larger "valence shell", owing to its larger size, than the smaller oxygen core O^{+6} (not shown, on the left).

O^{+6} in hypothetical OF_6 would, in the terminology of ionic models of matter, "rattle", settle "off center", favoring, thereby, ejection of F atoms, as F_2, and coordination by O^{+6} of lone pairs, yielding OF_2.

One can say, of course, that, nonetheless, the two cores have the same "valence manifold", defined in terms *s*- and *p*-orbitals available for bonding, and that bonding in SF_6 occurs via three-center/four-electron bonds, without any use of sulfur 3d orbitals in bonding sp^3d^2 hybrids (or even without use of sulfur 3s orbitals!).

In fact, a *completely ionic model of* SF_6, $S^{+6}(F^-)_6$, with tetrahedral charge distributions for the six F⁻ ligands, yields a reasonable, ball-park figure for the molecule's "lattice energy". Atomic cores such as O^{+6} and S^{+6}, it may be noted, may be viewed as small cations and cations such as Na^+ and Ca^{+2} as large atomic cores.

The statement that sulfur obeys the Octet Rule in "hypervalent" SF_6 uses (without notice or apology) the phrase "Octet Rule" in a *different sense* than Lewis did, to mean "number of sulfur atomic orbitals used in a theoretical description of the bonding" rather than to mean, more physically, after Lewis, "number of electrons in sulfur's valence shell". A question remains, however: Why doesn't oxygen exhibit "hypervalency"? The usual answer: 3c/4e bonding requires a large electronegativity difference between the central atom and its ligands. That electronegativity difference is relatively small, however, in the case of the triply-bonded nitrogen ligand of NSF_3. For the methyl ligands in Me_2SO, electron-pair coordination number for S^{+6} of 5, the sign of the electronegativity difference is reversed.

Gil Haight used to say that he illustrated all of chemistry in his general chemistry classes with "H-two plus O-two gives H-two-O". Similarly here. Many topics in chemistry and physics — among them atomic theory, descriptive inorganic chemistry, structural inorganic chemistry, atomic spectroscopy, quantum mechanics, and theories of the chemical bond — may influence location in a periodic table of an element such as hydrogen. Above lithium? Above fluorine? Above carbon? Above an entire table? Each location has its merits, and its advocates.

The purpose of this report is to report on the merits of H/Li — chiefly in conjunction with the merits of its companion, He/Be. One assignment without the other one is less interesting than the two assignments together, wherever one places H and He: above Li and Be (this report); above F and Ne (the author's preferred location for nearly four decades); or above the entire table (Jensen).

40. Ramsay, W. "An Undiscovered Gas," *Nature*, **56**, 378-382 (1897). Reprinted in (5), pp345-349. Noting that the difference of 36 in atomic weights of He and Ar [extremes of a *tertiary* triad (**94**)] is approximately equal to the difference in atomic weights of F and Mn, O and Cr, N and V, and C and Ti [extremes of secondary triads (**94**)], Ramsay predicted correctly (via two compensating wrong reasons) existence of an undiscovered inert gas with an atomic weight of 20. The history of the Periodic System includes several right results obtained for the wrong reasons — and many wrong results obtained by faulty use of correct facts.

41. Mendeleev (19), Volume II, p31. "[T]he history of the science of matter [in all its particularities], that is of chemistry, inevitably compels us to recognize the indestructibility not only of matter but also of the chemical elements," wrote Mendeleev in a famous footnote. "Therefore the thought involuntarily arises that there must be some bond of union between mass and the chemical elements; and as the mass of a substance is ultimately expressed . . . in the atom, a functional dependence should exist and be discoverable between individual properties of elements and their atomic weights. . . So I began to look about and write down the elements with their atomic weights and typical properties . . . on separate cards, and this soon convinced me that the properties of the elements are in periodic dependence on their atomic weights . . ."

42. Sanderson, R. T. "A Rational Periodic Table," *J. Chem. Ed.*, **41**, 181-189 (1964). Rationality exists in the mind of the reasoner. In the present author's mind, the LSPT offers more to reason

about than Sanderson's table does. Sanderson is "impatient with the fundamentally meaningless wide separation in the chart between beryllium and boron and magnesium and aluminum. [Also,] inclusion of zinc, cadmium, and mercury among the transition elements has always been disturbing." Recognizing that "The purpose of a periodic table may vary, and its design will vary appropriately," Sanderson suggests, nonetheless, as a general principle of periodic table design, the popular –

Principle of Similarity and Adjacency
"First and foremost we want similar elements grounded together."

"Similar" in what respect? As atoms? As simple substances? Or both? Sometimes one? And sometimes the other? Mg appears above both Ca and Zn, Y above both La and Lu, H above C, and, of course, He above Ne. The footnoted (and truncated) lanthanide and actinide series are spaced apart somewhat, owing to the relatively high maximum states of oxidation of elements 91 – 97.

Sanderson's "Rational Periodic Table" and the Left-Step Periodic Table are the north and south poles, so to speak, of tabular expression of, at one pole, all of Periodicity's arithmetical regularities and, at the other pole, graphic expression, via adjacency, of several important similarities among elements that are not adjacent to each other in the LSPT. Sanderson's table is a striking illustration of the principle that -

For every gain in tabular expression of chemical relationships, beyond those exhibited by Periodic Tables of no irregularities, there's a loss in regularity.

Sanderson's table "does a superior job of emphasizing the fact that the elements fall into four distinct electronic blocks," writes Jensen (24). One of Sanderson's blocks, his "F-block", is the same as the LSPT's *f*-block. His "D-block" has, however, only 9 columns, because the zinc Group (owing to its Group valence of 2) appears in a "P-block" (beneath Mg). Sanderson's table merges the LSPT's *s*- and *p*-blocks into a single block with $2 + 6 + 1 = 9$ members. H and He appear in a separate "S-block".

Sanderson's column labels are essentially the same as the "natural" column labels cited in Section **94**. For the carbon Group's "natural" label IVp Sanderson writes P4. The H and He (single-member) Groups have the labels S1 and S2, the Neon Group P8. In Sanderson's "Rational" Table, He and Ne are not in the same Group. Since the alkali and alkaline earth metals are in the same block as the other "main group elements", whose labels run from P3 to P8, Sanderson gives them the Group labels P1 and P2! He stretches the Be and Mg enclosures in his table to twice the width of the other elements' squares, so as to line up beneath them two columns of elements: Ca Sr Ba Ra and, to their right, Zn Cd Hg 112. Beneath Sc and Y in Sanderson's table are Lu and No.

On the whole, Sanderson's table is attractive, to outer and inner eyes. It's only 14 columns wide (the width of the F-Block), so it fits nicely on a page. It has the disadvantage that periods' rows are not lined up horizontally and, as Jensen notes, lost (when compared to a step-pyramid table) is "the ability of explicitly indicating secondary and tertiary relationships by means of broken and dotted lines." Losses, when compared to the LSPT, include absence of: indications of Periodicity's arithmetical regularities; adherence to a Correspondence with atomic physics; and ability to provide a ground for seeing Periodicity's block-to-block trends.

It's not difficult to create Sanderson's table from the LSPT. Place the latter's blocks one above the other, *f*-block at the bottom. Join the *s*-block to the left side of the *p*-block. Justify the blocks at the left. Double the width, rightward, of the Be and Mg enclosures. Move the Zinc Group to a position beneath Mg. Elevate and center above the *s/p*-block H and He. Add guide lines to indicate the relation of the (slightly truncated) *d*-block to the *s/p*-block and the *f*-block to the *d*-block. The transformations are, of course, reversible. By using Sanderson's table as it was designed to be used, with beginners, teachers are not exposing their students to something that might later have to be unlearned — particularly the "fact" that helium is a congener of neon.

43. Jensen, W., private communication, in the form of a step-pyramid periodic table that features a primary kinship tie-line between Mg and Zn. Zn is the only element in the table associated with *two* such tie-lines. Their translation into a block-style periodic table in which primary kinships are indicated by verticality yields Sanderson's periodic table (42).

44. Deming, Richard L.; Allred, A. L.; Dahl, Alan R.; Herlinger, Albert W. and Kestner, Mark O. "Tripositive Mercury. Low temperature oxidation of 1,4,8,11-tetraazacyclotetradecane mercury(II) tetrafluoroborate," *J. Am. Chem. Soc.,* **98**(14), 4132-7 (1976).

45. Weinhold, F. and Landis, C. R. *Valency and Bonding: A Natural Bond Orbital Donor-Acceptor Perspective,* Cambridge Univ. Press, London, 2005.

46. Bent, H. A. "Reflections on *Valency and Bonding"* by Weinhold and Landis, May-June 2003, privately circulated, pp6-7.

47. Seaborg, G. T. *Man-Made Transuranium Elements,* Prentice-Hall, Englewood Cliffs, N. J. 1963, Chapter 3; also: "The Transuranium Elements," *Science,* **104**, 379 (1946) and "Plutonium and Other Transuranium Elements, *Chem. Eng. News,* **25**, 358 (1947). In an article on "Prospects for Further Considerable Extension of the Periodic Table," *J. Chem. Ed.,* **46**, 626-634 (1969), Seaborg speaks of a "super-actinide *series* . . . postulated to contain 32 elements, whereas the lanthanides and actinide series each contain 14 elements" (emphasis added). Seaborg's figures leave the present author in doubt as to whether he considered the first 18 "super-actinides" to be the first row of a new block and the next 14 members congeners of lanthanides and actinides or whether he considered that all the elements — lanthanides, actinides, and "super-actinides" — are, or would be, members of special "series". In each of Seaborg's block-type periodic tables that the author has seen, the lanthanides and actinides appear as vertically separated "series". The phrase "*f*-block" does not appear in his articles, essays, reviews, and books on the actinides.

48. Sidgwick, N. V. *The Electronic Theory of Valency,* Oxford University Press, London, 1927. Sidgwick uses "a slightly modified form of Mendeleev's table" (Figure 23). His chief departures are removal of the latter's chess board character in its second period and *"inclusion of the whole of the rare earth metals (La 57 – Lu 71) in Group III A"* (emphasis added). He considers that "Hydrogen may be included either in group I or group VII: in either case, it is so peculiar as to require special treatment." It is a moot question whether considering hydrogen-above-lithium to be striking confirmation of a block-to-block trend in "first-element distinctiveness" is adequate "special treatment". The shape of the step-pyramid periodic table, *s*-elements on the left, places hydrogen with helium in a "block" of their own.

49. Paneth (33), p49. Cf. the Mendeleev-type "short-form" Table, p36, and the Conventional-type Periodic Table, p46. Paneth's reluctance, and that of many inorganic chemists of the pre-World War II era, and their students, to move thorium and tungsten in periodic tables from where Mendeleev located them, as congeners of, respectively, hafnium and tungsten, to the second row of an *f*-block is similar to reluctance to move helium from its conventional location above neon. In both instances a strong, if not close, kinship (secondary kinships in the cases of Th and U; a tertiary kinship in the case of He) is replaced by a weaker, but closer (primary), kinship (**83**).

50. Greenwood and Earnshaw (1), p1423. The elements "Z = 58-71" [sic] are called "this group".

51. Holleman, A. F. and Wiberg, E. *Inorganic Chemistry*, Academic Press, New York, 34th English edition, 1995, p1645. Cited is a "top element" rule for elements at the top of "main groups (Li, Be, B, C, N, O, F)". Not mentioned is the block-to-block trend in its magnitude. The volume's index provides an excellent guide to discussions of secondary kinships (**79**), of type pd, under the heading **periodic system** and the subheadings "comparison of main and transition group" III, IV, V, VI, and VII. Noteworthy, also, are updates, so to speak, of Mendeleev's "cards" (41). Given in three useful tables are numerical values for leading properties of the elements in (for convenient comparison) the format of the Periodic System's blocks.

52. Loening, K. L. "Recommended Format for the Periodic Table of the Elements," *J. Chem. Ed.*, **61**, 137 (1984). The top line below lists column labels recommended in 1984 by the American Chemical Society Committee on Nomenclature.

1	2	3d	4d	5d	6d	7d	8d	9d	10d	11d	12d	13	14	15	16	17	18
1	2	3d	4d	5d	6d	7d	8d	9d	10d	11d	12d	3p	4p	5p	6p	7p	8p
Li	Be											B	C	N	O	F	Ne
Na	Mg											Al	Si	P	S	Cl	Ar
K	Ca	Sc	Ti	V	Cr	Mn	Fe	Co	Ni	Cu	Zn	Ga	Ge	As	Se	Br	Kr
Rb	Sr	Y	Zr	Nb	Mo	Tc	Ru	Rh	Pd	Ag	Cd	In	Sn	Sb	Te	I	Xe
Fr	Ba	Lu	Hf	Ta	W	Re	Os	Ir	Pt	Au	Hg	Tl	Pb	Bi	Po	At	Rn

The ACS-recommended labels are almost identical with the "natural" labels (**94**) immediately beneath them, obtained by changing the "prefix" 1 of 13-18 to a suffix "p". In the natural scheme, the ACS's "3f" for the lanthanides and actinides (not shown) becomes 3f-16f. The change from

18 to 8p assumes, of course, that helium is not located above neon.

53. In *Principles*, Vol. II, p104, Mendeleev remarks that a "whole series" [sic] of elements "have long been classed under a special group". Then, uncharacteristically, he turns the discussion of the rare earths over to "Professor B. F. Brauner, of Prague, [who] has at my request written a special description of them for this book". Brauner concludes his description, written in 1902, with the remark that the rare earths are "difficult to arrange" in the periodic system but, "just as in [Mendeleev's] eighth group, four elements occupy one place in the system, so, also the rare earths form a node . . . and occupy [one] position". The rare earths' "node" expresses graphically and succinctly several leading facts about the rare earths, not hinted at by, e.g., an unembellished LSPT — at the expense, however, of Periodicity's overall regularities.

54. Hesse, M. B. *Forces and Fields,* Philosophical Library, New York, 1961, Chapter I: "The Logical Status of Theories: *Models."* The LSPT may be viewed as a model of the Periodic Law. Hesse notes, e.g., that a model "provides the context of natural expectations". [The LSPT raised expectations that led to most of the present report.] Also, models [like periodic tables] have been "devices which were essential [for ordinary intellects] for rendering [the Periodic Law] intelligible and testable". Additionally, a satisfactory model suggests analogies ["atomanalogies"] and raises questions [What, e.g., are the physical reasons for late occupancy of high-ℓ orbitals?]. Also, models can be modified [a periodic table annotated: e.g., with column labels] in ways that a formal theory, or verbal statement of a law, cannot. And, of course, to be right, i.e., particularly useful, a model [or periodic table] must be wanting, in some respects [e.g., in satisfying all four of the Construction Conventions cited in Figure 1 and, simultaneously, in giving graphic expression to all of Periodicity's different types of kinships], else it would be the thing itself.

"No single model [or periodic table] . . . is adequate to explain [or to exhibit all the dimensions of] the phenomena of the atomic domain [and its creation of Chemical Periodicity]", concludes Hesse. Consequently, in using a model [or periodic table] one must, in the words of a popular song of the 1940s, "accent the positive [the LSPT's regularities], diminuate the negative [unfamiliar block order]", and use complementary models [and tables] to express additional features of Chemical Periodicity.

55. Shchukarev (32), p229. Secondary periodicity is the 16th of sixteen "well-defined specific features of atoms" that Shchukarev lists as illuminated and unified by the concept of "kainosymmetry" (32). Cited are two *chemical* examples: "the toxicity of some arsenic compounds, and the absence of this property in analogous compounds of phosphorus and antimony; and the ability of selenic acid to bring metallic gold into solution, and the absence of this property in sulfuric and telluric acids".

56. Allen, L. C. "Electronegativity Is the Average One-Electron Energy of the Valence-Shell Electrons in Ground-State Free Atoms," *J. Am. Chem. Soc.*, **111**, 9003-914 (1989). Allen's "spectroscopic" elecronegativities are similar to those of Pauling and Allred & Rochow, with, for the present study, one particularly significant difference: Allen computes *electronegativities*

for the noble gases (also provided by Mulliken's definition, which yields similar results). Allen's values are:

Xe **2.6**   Kr **3.0**   Ar **3.2**   Ne **4.8**   **He** 4.2

Clearly it is the case that –

> *According to Allen's electronegativities, and Mendeleev's Rule of First-Element Distinctiveness, helium is not a satisfactory first-element for the Noble Gas Group, whereas neon is.*

Is it the case, however, that helium is a satisfactory first-element, according to Allen's electronegativities, for the Alkaline Earth Group? The "atomanalogy" H : Li :: He : Be yields for the electronegativity of helium [He = Be(H/Li)] the value 4.0, within five percent of Allen's value. As in the case of first-stage ionization energies, Allen's electronegativities support in three ways the change He/Ne to He/Be. His electronegativities illustrate, additionally, another use of the LSPT. Secondary periodicity's kinks for the s-block's Groups in plots of electronegativities against ordinal numbers of the sfdp periods of the Conventional Periodic Table are out of phase with those of the p-block. Plots against ordinal numbers of the fdps periods of the LSPT bring the blocks' kinks into phase with each other (Figure 44). In that sense, the ordinal numbers P_{fdps} are "natural". Generated one way, by regularities of primary periodicity, they assist in another way, in expression of regularities of secondary periodicity.

57. Mazurs (3), inside of the front cover. This periodic table is both perfectly symmetrical and perfectly regular. That perfection comes, however, at a high price. Although all members of a Group are in the same column, not all elements in a column are members of the same Group. Violated by Mazur's table is the cardinal construction convention that elements are in vertical alignment if and only if they are congeners. (In, e.g., the Li column of Figure 48 are, in addition to Li's congeners, all members of the N, Mn, and Eu Groups.) The table's horizontal coordinate has no chemical or physical significance. Its mathematical significance lies in the statement that the ordinal number within a row of the element immediately to the left of the table's center line is $2\ell + 1$.

58. Gorski, A. "Atomic Core Based Periodic System of Elements. A Contradictory Approach to the Arbitrary 1-18 Group Numbering in the Long Form of Atoms Based Periodic Table," *Polish J. Chem.*, **79**, 1435-1443 (2005). Figure 51 is similar to Gorski's figure titled "The block version of triparametric ["tridimensional"] periodic chart of atomic cores". Gorski's periods start, however, with the alkali metals. And he considers zinc to be a congener of magnesium.

59. Paneth (33), p49. Paneth believed that "the similarity between the elements of the *so-called actinide series* [emphasis added] is not nearly great enough to justify [Seaborg's] comparison with the rare-earth group [sic], and that it would be a mistake to obscure the obvious correspondence [indicated, e.g., by verticality in a periodic table] between [following in the footsteps of Mendeleev] lanthanum and actinium, thorium and hafnium, protoactinium and tantalum, uranium and tungsten." Consistent graphic expression by adjacency in a periodic table

of "obvious correspondence" among elements requires, in addition to adjacencies corresponding to the correspondences cited by Paneth: (i) helium adjacent (or connected by a tie-line) to neon; (ii) hydrogen adjacent (or etc.) to both carbon and fluorine; (iii) magnesium adjacent to both zinc and calcium; (iv) Al, Si, P, S, and Cl adjacent to, respectively, both Ga and Sc, Ge and Ti, As and V, Se and Cr, and Br and Mn; (v) all the rare earths and corresponding actinides in a super-Group III; and (vi) blocks of metals adjacent to blocks of metals. Produced, in words of a chemist, are "*true* periodic tables".

A plot against chemical "truth" of tabular expressions of arithmetical regularities starts low for Dobereiner's triads, rises significantly for Newland's octaves, rises dramatically for Mendeleev's tables, reaches a maximum for Janet's Left-Step Table, and declines thereafter with graphic expression of additional chemical truths. Less [truth] may mean more [regularity]. The ordinate of the curve's point of maximum regularity — and its point of maximum correspondence with atomic physics — is the LSPT. Its corresponding abscissa is graphic expression of primary chemical kinships, nothing more, nothing less. Beyond primary kinships lie secondary, tertiary, quaternary, special, knight's move, and other kinds of kinships and additional information regarding elements — metal/nonmetal character, physical state at standard conditions, radioactivity, crystal structure, etc. — whose graphic representation in periodic tables adds clutter not clarity to expression of the essence of Chemical Periodicity. Once that essence is understood, however, through use of the LSPT, the enhanced understanding achieved with its aid might lead to enhanced appreciation and use of chemically more familiar, "truth"-enhanced-and-regularity-reduced periodic tables.

60. The author is indebted to his, at the time, ten-year-old grandson Alex Bent Weberg for that observation. Aware that a chemist may use more than one periodic table, Alex framed for a Christmas present for his grandfather a large rectangular sheet of polished steel on which he placed 110 embossed 1.5 inch wood squares backed with magnetic tape. The changing display hangs in a place of honor in his maternal grandparents' home. Illustrated is one of the attractive features of Chemical Periodicity: its modular character. Created is what, according to Mark Twain, makes horse-racing interesting: differences of opinion. Examples include: problem-element placement; column labels; definitions of chemical kinships and "transition metal"; significance of the main-Group/sub-Group and metal/nonmetal dichotomies; the relevance of the Rules of Similarity and Adjacency and First-Element Distinctiveness; status of the Madelung Rule; and whether or not there is, indeed, a "best" periodic table.

61. Whewell (8), Vol. II. "A true Theory is a fact, a Fact is a familiar theory" (p652). "[An] inductive proposition is a *Theory* with regard to the Facts which it includes, while it is to be looked upon as a *Fact* with respect to the higher generalizations in which it is included" (p94). "We commonly call our observations Facts, when we apply, without effort or consciousness, conceptions perfectly familiar to us" (p94). "[T]he opposition of Fact and Theory is untenable, and leads to endless perplexity and debate" (p94). "[W]e are often told that such a thing is a *Fact* and not a theory [or a *Theory* and not a fact], with all the emphasis which, in speaking or writing, tone or italics

or capitals can give. . . Before we can estimate the truth, or the value of the assertion, we must ask to whom it is a fact [or theory]? what habits of thought, what previous information does it imply to conceive the fact as a fact?" (p655).

62. Weeks, M. E. *Discovery of the Element,* Journal of Chemical Education, 6th edition, 1956, p786. "For more than a quarter of a century most spectroscopists doubted the existence of Lockyer's 'helium'."

63. Shchukarev (55), p235. "As is well known," states the author, "Mendeleev expected that the secret of the periodic law would probably be discovered through the theory of numbers". A case in point: $\ell = 0$ and $e = 2$ for helium, not, as for neon, $\ell = 1$ and $e = 6$. "That's pure numerology," opines a critic of the assignment He/Be, "not chemistry." "Numerology" has, indeed, played a leading role in the history of chemical thought. Gay-Lussac and Dalton discovered secrets of chemistry in integers, in their laws of combining volumes and multiple proportions. Frankland discovered laws of integral valence, Faraday a law of integers regarding electrochemical equivalents. And Mendeleev's Periodic Law introduced into chemistry three additional sets of integers: elements' ordinal numbers in the Classification; "odd" and "even" "series"; and Groups' ordinal numbers within their "series".

Chemistry was the first inductive science to feature integers. That chemistry's integers have received an interpretation in terms of the integers of a quantum mechanical model of atoms is perhaps not surprising. *The physical atom and the chemical atom are the same atom. And only one set of integers exists.*

The isoelectronic principle is another chemical "secret" regarding integers. It might appear to relate He to Ne, in that an alchemical addition of a proton to the nucleus of each atom yields the related ions Li^+ and Na^+. That's not necessarily support for the assignment He/Ne, however, in that only at He is an exit from a block, with increasing atomic number, followed by reentry to the same block.

64. Wittgenstein, L. *Philosophical Investigations,* 3rd edition, trans. G. E. M. Anscombe, Macmillan Publishing Co., In., New York, 1968, Section 531: "We speak of understanding a sentence [or a periodic table] in the sense in which it can be replaced by another which says the same; but also in the sense in which it cannot be replaced by any other." All periodic tables are topologically equivalent to each other (**100**); but only the LSPT satisfies the four Construction Conventions (**1**). They are, so to speak, its grammar, its *"essence"* (S371). "To invent a language [via a grammar] could mean to invent an instrument [such as a periodic table of perfect regularity] for a particular purpose [revelation of Periodicity's regularities] on the basis of the laws of nature [the Periodic Law]" (S492). The LSPT is, in Wittgenstein's phrase, *a perspicuous representation,* in that it "produces just that understanding which consists in 'seeing connections'" (S122). "If some one says 'If our language [or table] had not this grammar [or shape] it could not express these facts' — it should be asked what *'could'* means here" (S497). Might it mean [for the LSPT] easy expression of support for the proposition He/Be?

Of a proposition's critics, Wittgenstein asks: "What sort of information do you call a ground for such a belief? . . . In what kind of way do you expect to be convinced?" (S481). His answer (Part II, section xii): "[I]f anyone believes that certain concepts [e.g., He/Ne] are absolutely the correct ones, and that having different ones [He/Be] would mean not realizing something that we realize — then let him imagine certain very general facts of nature [e.g., effects of electron screening and penetration on energy levels of many-electron atoms] to be different from what we are used to [68], and the formation of concepts different from the usual ones will become intelligible to him."

65. Mendeleev (19), Vol. II, p517. Had Mendeleev known of XeO_4, he'd have had diminished grounds for a "zero group", hence perhaps no zero series, no presumed elements lighter than hydrogen, no "chemical conception of the ether", no consequent tarnishing of his reputation — and, consequently, perhaps no absence of a Nobel Prize?

66. Wittgenstein offers an explanation for existence of problematical propositions, such as: Helium is a Noble Gas. "[W]hen we are tempted in philosophy to count some quite useless [or almost quite useless] thing a proposition, that is often because *we have not considered its application sufficiently*" (S520, emphasis added). Application of the proposition He/Ne alters the character of two Groups in the Periodic System and, thereby, two blocks of Groups and invalidates all the generalizations and trends cited in this report. Falsification of any part of an axiomatic system falsifies the entire system.

67. Mendeleev (19), Vol. II, pp508. Mendeleev remarks, regarding prior studies of De Chancourtois and Newlands, "that they merely wanted the boldness necessary to place the whole question [of a classification of the chemical elements] at such a height that its reflection on the facts could be clearly seen" (p493). His remark applies literally to helium's height in the LSPT.

   "[B]oth thorium and uranium are of great importance to the periodic system," adds Mendeleev, "as they are the last members" (p503). Of greater importance to the system (now known to be endless) are its first two members.

68. Jensen, W. B. *FIN DE SIECLE CHEMISTRY 1400-2000. A Brief Survey of 600 Years of Chemical History for Students of Chemistry,* Oesper Collections in the History of Chemistry, 2003, p268.

69. Pais, A. *Niels Bohr's Times, In Physics, Philosophy, and Polity,* Clarenden Press, Oxford, 1991, pp208-9. In describing (p205) the excitement, anticipation, and frustration physicists felt regarding Bohr's reports of his studies of an electronic interpretation of the periodic table, Pais provides a rationale for the chemical capture of the LSPT sketched in Figure 1.

   Sommerfeld to Bohr (1921): "If, as it appears, you can reconstruct mathematically the numbers 2, 8, 18, . . . of the elements in the periods, that is in fact fulfillment of the boldest hopes in physics."

   Rutherford to Bohr (1921): "It is very difficult to form an idea of how you arrive at your conclusions. Everyone is eager to know whether you can fix the 'rings of electrons' by the correspondence principle or whether you have recourse to the chemical facts to do so."

Frank to Bohr (1922): "The curiosity of the local physicists to get to know your methods mathematically as well is tremendous."

Neither then nor later, writes Pais, would their curiosity be satisfied.

Heisenberg, in his book *Physics and Beyond*, 1971, writes that "We could clearly sense that [Bohr] had reached his results not so much by calculations and demonstrations as by intuition . . ."

Asked by a chemist why he placed lithium's third electron in a second, outer orbit, Bohr is said to have replied (in a moment of frankness): "The facts of chemistry require it."

In hindsight, the ionization energies of the first three elements, 13.6, 24.6, and **5.4 (!)** eV virtually demand, by induction, a Principle of Spatial Exclusion:

> *Two but no more than two electrons can be at the same place at*
> *the same time.*

The Spatial Exclusion Principle is consistent with a union of Lewis's conclusion regarding the valence-strokes of classical bond diagrams — that they represent *two* electrons — with a seldom-stated theorem: In satisfactory bond diagrams –

> *Valence-strokes never cross each other.*

Ruled out, e.g., is a planar centric formula for benzene:

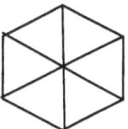

The current interpretation of the Principal of Spatial Exclusion assigns to electrons "a classically nondescribable two-valuedness" (Pauli), called "spin", value plus or minus 1/2, and states:

> *Two electrons of the same spin cannot be at the*
> *same place at the same time.*

The prohibition is a mathematically rigorous implication of Born's interpretation of the physical significance of wave functions that, after Heisenberg, are antisymmetric to exchanges of coordinates of identical, spin 1/2 "fermions".

How large "in practice" valence-shell electrons' van der Waals-like "fermi holes" are can be determined by a simple experiment. Replace valence-strokes of bond-diagrams by impenetrable, valence-shell-filling valence-spheres and see what structures emerge. As one might expect from the close-packed arrangement of four cannon balls and carbon atoms' famous "tetrahedral" character, locations in valence-sphere packing models of sites for atomic cores correspond to observed molecular shapes for molecules that can be represented satisfactorily by valence-stroke diagrams.

Provided, in passing, is an interpretation of Periodicity's first two magic numbers: 2 and 8.

70. Bochner, S. *The Role of Mathematics in the Rise of Science,* Princeton University Press, Princeton, NJ, 1966, p237-8. "The first significant 'universal' history of science, still very readable," writes Bochner (p69), "came several decades after 1800, and it was due to the author of diverse works, William Whewell (1796-1866)". Faraday consulted Whewell regarding terms for new concepts. Best known are "cation" and "anion" — the central concepts, under the names "atomic core" and "electride ion" (or "electron-pair"), of electron-domain models of molecules (71). (Cf. Appendix XIII.)

71. Bent, H. A., "Ion-Packing Models of Covalent Compounds," *J. Chem. Ed.,* **45**, 768-778 (1968). *All substances,* it's concluded, *are ion-compounds.* Analogous terms for heteropolar and homopolar compounds include: crystal/molecule; cation/atomic core; anion/valence-shell electron pair; cation in a tetrahedral interstice/octet rule obeyed; bridging anion/bonding pair; corner-sharing (of coordination polyhedra)/single bond; edge-sharing/double bond; face-sharing/triple bond; cation-cation coulomb repulsion/Baeyer strain energy (owing to core-core repulsion); anion lattice/bond diagram. Pauling's First Rule of Crystal Chemistry, that in ionic compounds a coordinated polyhedron of anions is formed about each cation, becomes in "electride-ion" chemistry the rule that in molecules a coordinated polyhedron of electron-pairs is formed about each atomic core.

72. Bent, H. A. and Bent, H. E. "What do I remember? The Role of Lecture Demonstrations in Teaching Chemistry," *J. Chem. Ed.,* **57**, 609-618 (1980).

73. Lavoisier, A. *Elements of Chemistry,* Dover Publications, Inc., New York, 1965, Preface, pxvi. "[I]n commencing the study of a physical science we ought to form no idea but what is a necessary consequence, and an immediate effect, of an experiment or an observation." Of course, the notion that the idea that helium is a congener of neon "is a necessary consequence, and an immediate effect, of [the] observation" that helium and neon are both inert gases raises the question: What's to be included as part of an observation? Should one include in the helium-neon observation Mendeleev's remark regarding first-element distinctiveness? To paraphrase Lavoisier's remarks in his Preface's concluding paragraph (pxxxvii): Does helium-above-neon in periodic tables communicate to observers "that precision and accuracy [and comprehensiveness] which [scientists] have employed in their observations"?

74. Whitehead, A. N. *The Aims of Education and Other Essays,* The Free Press, New York, 1929, p37. "It should be the chief aim of a university professor to exhibit himself in his own true character—that is, as an ignorant man thinking, actively using his small share of knowledge." That happens in spades when demonstration-experiments don't "work" — that is to say, when they don't proceed as expected. Provided are excellent opportunities for a scientist to exhibit him or herself in one's own true character, as a person ignorant, initially, of the cause(s) of Nature's unexpected behavior, thinking, actively using one's knowledge, and aggressively pursuing opportunities for additional experiments and observations, in efforts to account for

what happened, in the firm belief that, since Nature always does her thing (so far as we know), there's a reason for everything. Audiences love it. "If there's one thing my kids love to do," said an enthusiastic middle school teacher following an all-school assembly in which a demonstration-experiment hadn't 'worked' (the first time), "it is to think."

Whitehead's book is a gold mine of wisdom. It should be required reading, regularly, like a bible, of all teachers, school administrators, state and federal agencies concerned with education, and self-styled "education presidents". His remark that "The best education is to be found in gaining the utmost information from the simplest apparatus" describes accurately the present author's experiences with the Left-Step Periodic Table. His observation that "The subjects pursued for the sake of a general education [in, e.g., inorganic chemistry] are special subjects [Periodicity] specially studied [with the LSPT]" aptly describes studies whose outcomes are reported in this report. His commandment "What you teach, teach thoroughly" accounts for the report's length.

75. Zinsser, W. *On Writing Well: An Informal Guide to Writing Nonfiction,* Harper & Row, Publishers, New York, 1976, Chapter 14: Science. "The assignment I give is a seemingly primitive one. I just ask [students] to describe how something works. I don't care about a seductive lead or a surprise ending or any connecting devices. . . It's just a question of putting one sentence after another. The 'after', however, is unusually important. Nowhere else must you work so hard to write sentences that form a linear sequence. This is no place for fanciful leaps or implied truths. Fact and deduction are the ruling family. . . I have found it to be a breakthrough assignment for many students who just couldn't disentangle themselves from vagueness, cluttered, and disorderly thinking.

NOTE ADDED IN PROOF: A Striking Example of a Particular Statement of a General Feature of the Periodic System.

7.12.06

Dear Roald,

Thank you for the copy of a copy of the manuscript by Philip J. Stewart that you received from Oliver Sachs, on "Tables and Spirals". It calls to mind four statements about periodic tables.

1. No periodic table has all the desirable features periodic tables might have.

2. No periodic table is superior to all other periodic tables in all respects.

3. No *best* periodic table exists, since the choice depends on use(s) to which the table is put.

4. Because the Periodic Law is one of the central doctrines of the Central Science, tabular expressions of it have many uses. There exist, consequently, many periodic tables.

Stewart's manuscript suggests yet another way to say pretty much the same thing:

For any graphic representation of Chemical Periodicity it is possible to select criteria of suitability that make that particular representation the most suitable one.

The Table Construction Conventions cited at the outset of this report make the Left-Step Table the most suitable one. Similarly -

- A desire to display secondary chemical kinships alongside primary kinships, exhibited vertically, makes Mendeleev's "Short Form" Table the most suitable one.

- A desire to exhibit secondary kinships in a table of gapless periods makes a Step-Pyramid Table the most suitable one.

- A desire to exhibit the distinctiveness of the *s*-block in a two-dimensional arrangement of atoms makes the Conventional Periodic Table the most suitable one.

- A desire to exhibit vertically the order of filling of atomic subshells in Bohr's Aufbau Process makes the Right-Step Table the most suitable one.

- A desire to exhibit graphically in a table of no irregularities Chemical Periodicity's three-dimensional character makes the arrangement of atoms in Pℓe space the most suitable arrangement. And -

- A desire, in Stewart's words, "to evoke wonder" in unsophisticated viewers makes his "Chemical [Spiral] Galaxy", with its starry background, the neutron as his spiral's first "element", and reference to neutron stars, the most suitable arrangement, in Stewart's view. With its placement of hydrogen in the Carbon Group and helium in the Neon Group, Stewart's "Chemical Galaxy" violates, however, the Orbital Radial Nodal Surface Rule for five of the Periodic System's thirty-two Groups.

Wonder for the outer eye is not necessarily the same thing as wonder for the inner eye.

Except for n, H, and He, Stewart's spiral is Janet's spiral form of his Left-Step Table, in which the Table's left end is spiraled around, clockwise, until its first two elements 57 and 89 (La and Ac of the f-block) are adjacent to the two elements 56 and 88 (Ba and Ra of the *s*-block). Arcs connect the remainder of the alkaline earth metals, and He, to their immediate successors in the Periodic System.

Best Regards,

Henry

Final Words

"Sometimes the hardest things to see," artists say,
"are the things that stare us in the face."

The *s*-block is unique. The Periodic System's most metallic element lies at its lower left corner. The System's most distinctive element lies at its upper right corner. Consequently, to exhibit "across the board" trends in metallic character, locate the *s*-block on the left-hand-side of periodic tables. To exhibit trends in distinctiveness, locate it on the right-hand-side.

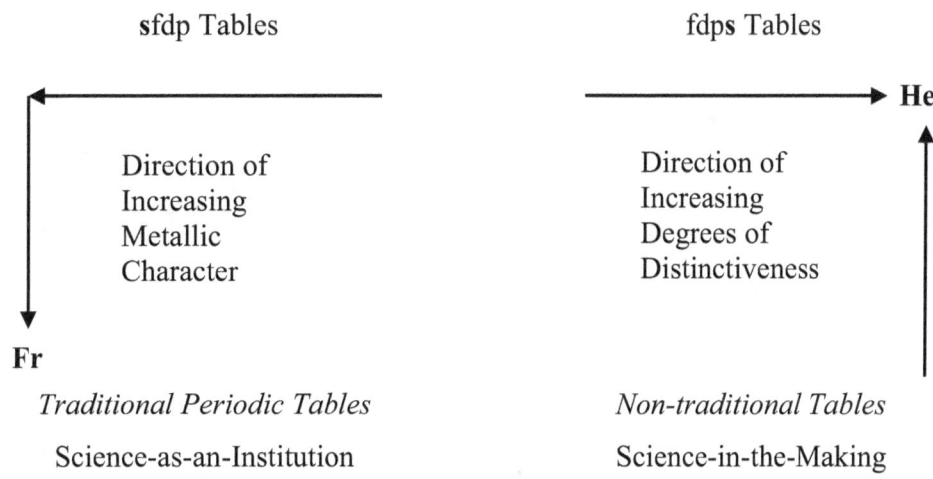

"Final Words" (above) are not, hopefully, final words for Periodic System Systematics. "[I]t needs not only new applications," to quote Mendeleev's concluding remarks about the Periodic Law in his Faraday Lecture of 1889, "but also improvements, further development, and plenty of fresh energy."

Henry Bent
July 2006

The End

of

Brian's final gift to chemical thought.*

---

* After Brian died his uncle Bob said to his father "Just do something Brian would be proud of."
This report may come as close to that as Brian's father can come.

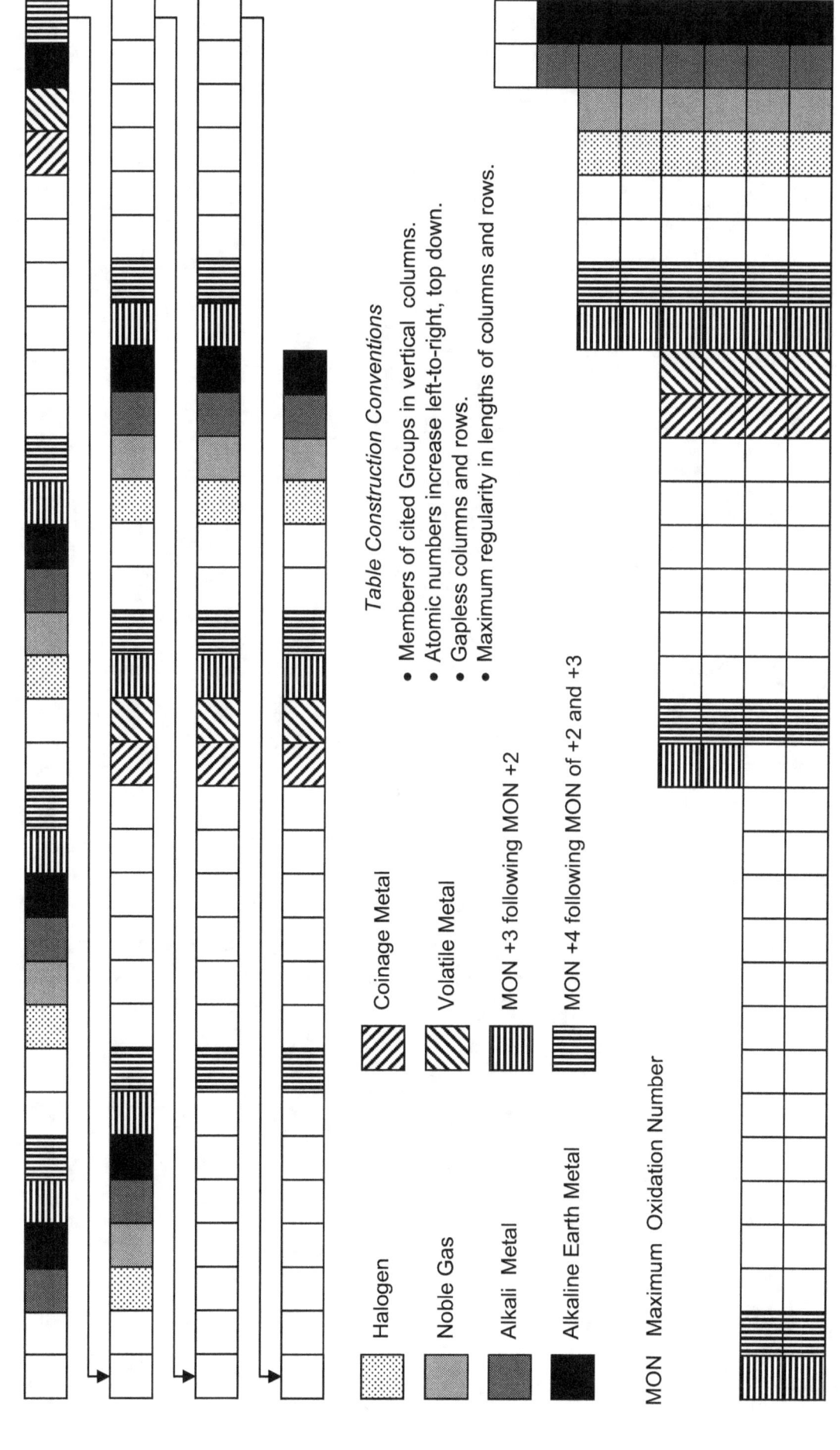

The elements, if arranged according to their atomic weights, exhibit an evident *periodicity* of properties.

DIMITRI MENDELEEV, 1869

Table Construction Conventions
- Members of cited Groups in vertical columns.
- Atomic numbers increase left-to-right, top down.
- Gapless columns and rows.
- Maximum regularity in lengths of columns and rows.

Halogen

Noble Gas

Alkali Metal

Alkaline Earth Metal

Coinage Metal

Volatile Metal

MON +3 following MON +2

MON +4 following MON of +2 and +3

MON   Maximum Oxidation Number